ACTIVIST BIOLOGY

LATIN AMERICAN LANDSCAPES

REGINA HORTA DUARTE

ACTIVIST BIOLOGY

The National Museum, Politics, and Nation Building in Brazil

TRANSLATED BY
DIANE GROSKLAUS WHITTY

THE UNIVERSITY OF ARIZONA PRESS
TUCSON

The University of Arizona Press
www.uapress.arizona.edu

Printed in the United States of America
21 20 19 18 17 16 6 5 4 3 2 1

ISBN-13: 978-0-8165-3201-8 (cloth)

Cover design by Miriam Warren
Cover photo of *Agrias phalcidon phalcidon ab. atromarginata*, courtesy of Notafly/Wikimedia

Originally published as *A Biologia Militante: O Museu Nacional, especialização científica, divulgação do conhecimento e práticas científicas no Brasil (1926–1945)* by Editora UFMG, 2010.

Publication of this book is made possible in part by the proceeds of a permanent endowment created with the assistance of a Challenge Grant from the National Endowment for the Humanities, a federal agency.

Library of Congress Cataloging-in-Publication Data
Names: Duarte, Regina Horta, author. | Expanded version of (work): Duarte, Regina Horta. A biologia militante.
Title: Activist biology : the National Museum, politics, and nation building in Brazil / Regina Horta Duarte ; translated by Diane Grosklaus Whitty.
Other titles: Latin American landscapes.
Description: Tucson : The University of Arizona Press, 2016. | Series: Latin American landscapes | Includes bibliographical references and index. | Revised and expanded version of A biologia militante : o Museu Nacional, especialização científica, divulgação do conhecimento e práticas científicas no Brasil 1926–1945.
Identifiers: LCCN 2016007486 | ISBN 9780816532018 (cloth : alk. paper)
Subjects: LCSH: Biology—Brazil—History—20th century. | Museu Nacional (Brazil)
Classification: LCC QH305.2.B6 D8213 2016 | DDC 578.0981—dc23 LC record available at https://lccn.loc.gov/2016007486

♾ This paper meets the requirements of ANSI/NISO Z39.48-1992 (Permanence of Paper).

For Dora and Manoel,
beloved parents

For Tom,
my oasis

CONTENTS

ILLUSTRATIONS

PREFACE AND ACKNOWLEDGMENTS

THIS BOOK IS A REVISED AND EXPANDED VERSION OF *A Biologia Militante: O Museu Nacional, especialização científica, divulgação do conhecimento e práticas científicas no Brasil (1926–1945)*, published in 2010 by Editora UFMG. My overriding objective in undertaking these revisions was to bring concepts, argumentation, and the narrative itself into sharper focus. New primary and secondary sources were incorporated as I refined my conceptualizations and analyses, lending greater cohesiveness and strength to the book's overall argument. I expanded on my discussions of the historical, political, and social context, especially regarding the National Museum, the First Republic (1889–1930), and the Vargas era. I also added a timeline and a chronological table that delineates the phases of the Republic addressed in these pages—information that will be especially helpful to the non-Brazilian public.

Expressing gratitude should be more than an act of mere formality, for it derives from the certainty that we are not alone. Over the years, I have relied on the immense support of many people, and my list of acknowledgments could be endless. I will begin with the public institutions in Brazil that welcomed and supported me: the Escola Estadual José Bonifácio, where I learned to read; the Universidade Federal de Minas Gerais (UFMG); the Universidade Estadual de Campinas; the Conselho Nacional de Desenvolvimento Científico e Tecnológico (CNPq); and the Fundação de Amparo à Pesquisa do Estado de Minas Gerais. I am especially grateful to the CNPq, which funded this research, and to the Instituto de Estudos Avançados Transdisciplinares, where I had the privilege, as a resident scholar in 2008, of spending a period of study and intellectual fellowship that was vital to this project.

I received invaluable help from Vilma Carvalho de Souza, of the Biblioteca Prof. Antônio Luiz Paixão; from the whole team at the Seção de Memória e Arquivo do Museu Nacional; from Hermínia Ferreira, of the Academia Brasileira de Ciências; from Juliana Amorim, of the Arquivo Múcio Leão at the Academia Brasileira de Letras; and from the staff at the Museo Nacional de Historia Natural in Montevideo. My thanks go as well to my colleagues and students and to the staff at the UFMG Department of History. My participation in the Coleção Brasiliana research group, led by Eliana Dutra, greatly influenced my decision to research this particular topic—to all my *brasilianos* colleagues, a huge *abraço*.

Lise Sedrez and Chris Boyer gave my work a warm welcome in the University of Arizona Press's Latin American Landscapes series. Kristen A. Buckles, also of the University of Arizona Press, was extremely zealous in her guidance of the editorial process. The reviewers of the original manuscript offered valuable suggestions, and I am deeply indebted to them for their thoughtful readings. Diane Grosklaus Whitty translated the book, and I learned much from her about the elegance and exactitude of the English language. Her exacting, tireless, and skillful work greatly enhanced the final text. Our joint labors and discussions of details of the English version behind us, I realize I have a fine new friend and cohort.

Any words of thanks to my family are far from enough: to my mother and father (for whom I'll always be their little girl); to my brothers and sisters, nieces and nephews. To Tom, my lifelong love, who is always there for me with words of encouragement and affection. Life with my sons, Antonio and Manoel, has been a perpetual joy. They taught me to like Radiohead, Pearl Jam, and Nirvana and enjoy every last episode of *Breaking Bad* and *Game of Thrones* (guilt free), while I in turn introduced them to those inimitable old-timers: Jimi Hendrix, Janis Joplin, and David Bowie. To you, my little boys—now so grown up and independent, off conquering your own worlds—I give thanks from the bottom of my heart.

TRANSLATOR'S NOTE

THE BOOK YOU ARE ABOUT TO READ is far from a straightforward English version of the original Portuguese. Regina Horta Duarte made a series of revisions to the manuscript before delivering it to me, as she explains in the preface. Thanks to the rich collaborative relationship the author and I developed, I had the liberty to suggest other adjustments, always reviewed by Regina, for example, the rearrangement and relocation of paragraphs here and there, very minor cuts, a few notes to point out errors detected in primary sources. Citations of Foucault and Mayr in Portuguese were replaced by their classic English versions (a task Regina graciously insisted was hers), ensuring that the present audience has access to the most appropriate bibliography.

The outcome is a highly domesticated text. In my view, the content itself abounds in foreignness, and I could best do my job by smoothing the reading process as much as possible. This at times led to what we might call "activist translation," where I retained a semblance of invisibility while posing as the author. For instance, the cultural and historical gaps unnoticed by the Lusophone reader are filled in not by translator notes but by direct intervention in the text, and so "the place where the cross had first been planted" becomes "the place where the cross had first been planted by the Portuguese discoverers of Brazil on April 26, 1500." Similarly, the linguistic subtleties of Roquette-Pinto's distinction between *brasileiro* and *brasiliano* are explained as if in Regina's voice.

You will hear another constant voice, or set of voices: those of the scientists of the National Museum and their contemporaries. In these cases, I have stayed as faithful to the original as possible. This includes moments when Regina steps unobtrusively into the background to give her subjects the narrative stage, accounting for some uses of non–gender neutral patterns. The author and I were also in accord about respecting the historicity of proper names, which thus are spelled as they were in their day, despite any subsequent orthographic changes to Portuguese.

A hybrid approach has been applied to book, article, and film titles. Most are followed by literal parenthetical translations, but these have been omitted where it was felt a description would suffice. To facilitate the researcher's access to documents, original titles can be found in the notes when they do not appear in the text.

Lastly, all names of institutions, agencies, conferences, and so on appear in English, with their originals supplied in a list at the back of the book. A few common nouns have been retained in Portuguese. These are defined upon first occurrence in the text.

In closing, deepest thanks to my husband, and first reader, who saved me the embarrassment of inadvertently nailing a taxidermied jaguar carcass to a wall. Without Michael, no *oeuvre* is possible. My gratitude as well to skilled translator Kim Olson, who put on her copy editor's hat and went above and beyond to chip the rough edges off my prose. Lastly, my immeasurable thanks to Regina Horta Duarte, *gente finíssima*, who has graced my work life with her good humor, boundless patience, and, now, friendship. I am grateful for the privilege of sharing with English-speaking readers her nuanced analysis of three exemplary Brazilian scientists who endeavored to forge a new field and a new nation.

ACTIVIST BIOLOGY

INTRODUCTION

THE YEAR IS 2008. A GROUP OF TOURISTS HEADS out of São Luís, capital of Maranhão, where the beaches have been declared off-limits. On January 7, the local papers announced that researchers at the Federal University of Maranhão had detected fecal coliform levels twenty-five times above acceptable limits.

With the sun straight overhead, the vacationers point their car toward the town of Barreirinhas, not far from the promised paradise of Brazil's Lençóis Maranhenses, a national park since 1981. The road is nearly deserted, the landscape dotted by only a few villages, some scattered adobe houses, and a bar here and there. Large expanses of land have been burned off to give way to subsistence farming.

Their rental car starts acting up, and the tourists scramble for their guidebook, where they locate the lone gas station along the 100-mile stretch of road between them and the park. Worried they'll find themselves stuck in the middle of nowhere, the tourists press on—and breathe a heavy sigh of relief when they reach their unplanned stop. Their happiness, however, is short-lived. The gas station proves as forsaken as the rest of the area, and the attendants know nothing more than how to fill a tank.

A native bird hangs in a locked cage outside a humble house nearby. Three men have just finished their lunch and throw what little is left to two squalid mutts, followed eagerly by some pigs and a chicken. One of the men goes back to building a birdcage. A TV set hooked up to a satellite dish is playing a movie that stars Denzel Washington. The nearest public phone is almost nine miles away. Luckily, one of the travelers' cell phones picks up a signal and comes back to life, so they call the rental agency. Help will arrive in two hours.

While they wait, the tourists amble over to the nearby house, where lettering painted on an outside wall advertises a bar inside. It's closed, but the next-door neighbor is selling beverages and cookies. There's a pool table and some chairs on the porch. The family has an old jalopy, likely a sign of great prosperity in these parts. Every once in a while, dogs and chickens at their heels, a few curious children scurry in and out of the house to peer at the strangers. One of the tourists asks if there's a restroom she can use. Yes, around back. As she crosses the room, she sees men, women, and children crowded around another TV. The bathroom is outside, surrounded by chickens, pigs, and dogs. There's a porcelain toilet but it doesn't flush; a big bucket of dingy water and a small basin make up for it.

The other tourists—all men—are too embarrassed to breach the privacy of the home, so they go back to the gas station and ask to use the bathroom there. The fellow says they don't have one. "If you gotta take a dump, it's a little ways out back. But if you gotta pee, well, it's back over there too." He points to the woods behind the station. There, in the spot where the call of nature is answered, the visitors run into the same pigs that had been fighting over the lunchtime leftovers, now wallowing in a rather questionable puddle.

Help finally comes. The tank is filled, but the invoice will have to be issued in the town of Humberto de Campos, nine miles away, because the attendant doesn't know how to write.

The tourists continue on toward Barreirinhas, just outside the famous Lençóis Maranhenses and its dazzling display of undulating sand dunes ribboned with blue waters. The Lençóis lie in a national park that has no entrance gate and no supervision or form of control whatsoever. The lush plant life leading up to the dunes is marred by areas of burn-off and cropland. The park itself sits amid villages that lack safe drinking water, sewer systems, or garbage service. Its pathways are cluttered with plastic bags and bottles and beer cans routinely tossed away by tourists. Naked children with protruding bellies wander around among scruffy dogs and pigs, waving at the folks in passing cars, on their way from their luxury hotels or resorts to the region's magnificent (really?) tourist attractions.

THE NATIONAL MUSEUM

This was my own family's little adventure—but it was something that could have happened in many places in Brazil, where poverty, government neglect, unhealthy living conditions, illiteracy, consumerism, and tourism live side by side. When I saw the television set with its excellent reception and sharp images, blaring away in a lost corner of the country where life is so brutally harsh, it immediately brought to mind the history of a certain period of Brazil's National Museum. In the 1920s and 1930s, the scientists who worked at this institute in Rio de Janeiro hoped to transform it into a hub that would radiate knowledge to the farthest reaches of Brazil. During those years, the museum staff devoted itself tirelessly to re-creating the National Museum and staking claim to a new role for it. They couldn't begin to imagine television or satellite dishes, but they trusted in print, movies and radio, exhibits, and educational methodology as efficacious methods for disseminating the new knowledge and new practices that they were convinced would transform Brazil.

The National Museum already had a long history behind it by then. King Dom João VI of the United Kingdom of Portugal, Brazil, and the Algarves had founded it in 1818. His court had fled Lisbon shortly before the city was invaded by Napoleon's troops in 1808, and once settled safe and sound in Brazil, Dom João VI did his best to prepare Rio de Janeiro for its new status as the political and administrative center of the kingdom, a process that transported the seat of the European empire to the heart of the old Portuguese colony. The Royal Museum—as it was then known—emulated Old World museums by gathering collections representative of the entire globe. But the spotlight was on the Portuguese Empire, spread across the European, African, Asian, and South American continents. From its founding on, the museum played a decisive role in the development of natural history in Brazil.[1]

The establishment of the museum figured into a broader nineteenth-century trend around the world to set up natural history museums as "cathedrals to science." By 1910, there were some two thousand museums of its kind.[2] In Latin America, natural history museums enabled exchanges between naturalists while connecting different points of the globe. In cities like Rio de Janeiro, Buenos Aires, Santiago do Chile, Montevideo, Bogotá, and Caracas, new institutions continued to open their doors throughout the nineteenth century, almost always concomitant with processes of achieving independence and nation building. They were home to enlightened elites who combined their experience as locals with intellectual training in Europe, but they were also frequented by foreign naturalists eager to research the flora and fauna of South America. National governments wanted to undertake inventories of "their" nature and would often hire teams of foreigners to lend impetus to natural history.

The daily routine at nineteenth-century museums in Latin America reflected the challenges specific to the continent's historical context. Foreign scientific expeditions often took everything they gathered back to Europe, leaving nothing to the institutes that had welcomed and aided them, much to the discontent of local science communities. The piecemeal nature of local collections left Latin American naturalists at a tremendous disadvantage vis-à-vis their foreign peers, whose institutes boasted enviable collections. Latin American museums also had to cope with periodic political turmoil, which occasioned wild fluctuations in government funding and other support. As Nancy Stepan has said, a great deal of progress came thanks to the individual efforts of naturalists in the absence of any collective, institutionalized, stable climate. Teaching institutions emphasized a liberal arts education in a framework where there was no real way to train researchers in scientific work. There was a paucity of equipment and bibliographic material, scientists enjoyed little prestige, and agricultural and industrial modernization was not yet hardy enough to provide new sources of support for science.[3]

Many aspects of the history of these museums give nuance to the traditional view that naturalists working in Latin America were members of cloistered scientific communities.[4] Over the course of the nineteenth century, while Brazil's National Museum was becoming a place for public exhibits, it was also making room for new fields of knowledge in its various departments—like paleontology, anthropology, comparative anatomy and zoology, botany, mineralogy, geology, and archaeology—reflecting the institution's attention to research and its tendency to develop specialized fields. From 1876 to 1893, during what was known as "the golden age of the National Museum," the institution saw substantial changes under the direction of the naturalist Ladislau Netto. Its collections grew through exchange programs with European and Latin American counterparts and thanks to national expeditions financed by the imperial government. The old monarchical tradition of handpicking personnel by appointment was replaced by the requirement that new hires take qualifying exams on scientific topics. Foreigners like Charles Hartt, Fritz Müller, Hermann Von Ihering, Emílio Goeldi, Carl Schwacke, and Orville Derby were recruited and had plenty of opportunity for the rewarding exchange of experience and knowledge with Brazilian scholars.[5] The establishment of a laboratory for experimental physiology and the launching of a science journal in 1876 (*Arquivos do Museu Nacional*) energized the museum and cleared the path for its naturalist members to advance in their professionalization.[6] The institute's collaboration with the Brazilian presence at universal exhibitions was also important. The country wanted to make a place for itself on the world market and to be counted as a civilization in the tropics. It was not just its commercial interests that were at stake; so too were the exchange of scientific and technological know-how and interaction between the National Museum and foreign science institutes. No less important was the organization of Brazil's National Anthropological Exposition in 1892, where the exhibiting of hundreds of ethnographic objects fed the lively contemporary debate on race, people, and the Brazilian nation.[7]

Like other museums in Latin America—for example, the Argentina Museum of Natural Sciences (now the Bernardino Rivadávia Museum of Natural History), in Buenos Aires, headed by Hermann Burmeister—Brazil's National Museum experienced such profound changes during these years that it was almost like starting over. As naturalists reclassified nature, as knowledge grew more specialized, and as scientists and observers began relating to collections in new ways, these collections underwent extensive reorganization. In pursuing this new vision, the museum entered into the wider debate about "national being" and introduced a state "optic"—to use Andermann's term—of the items on display, thereby transforming a tour of the museum into a civics lesson.[8]

In 1889, the army, with the backing of the agro-exporting elites, toppled the monarchy, and Brazil became a republic. As much as civilian republican groups had hopes for a new democratic order, the institutions of the fledgling republic were predominantly individualist and liberal in nature, and most citizens were denied their political rights, since illiterates were prohibited from voting. Although slavery had been abolished under the monarchy, in 1888, the early decades of the republic witnessed no advances in civil and political rights; instead, it was an era of "exclusionary liberalism," or "oligarchical liberalism," marked by political accords between powerful elites, underwritten by fraudulent elections. The Constitution of 1891 delegated broad fiscal and administrative autonomy to the states and territories, benefiting the chief commodity-producing states, like coffee-rich São Paulo and the rubber centers of Pará and Amazonas. Under the influence of some republican sectors, the nation's charter also bore the imprint of positivism, translated into a complete separation of church and state and the absence of any official religion. The republic would recognize marriages, births, and burials as civil processes, and religious teaching would no longer be mandatory in schools.[9]

In the early years of the republic, the museum faced several hurdles. The new government abolished the post of traveling naturalist and demanded the daily physical presence of all researchers. In practical terms, this meant naturalists could not make research trips and instead had to stay in their offices. Some of the top staff left, Fritz Müller among them. A number of wealthier states, like São Paulo and Pará, opened their own natural history museums and managed to attract naturalists like Goeldi and Von Ihering. The federal government itself established applied research institutes, which became the country's first centers for biological research, such as the Oswaldo Cruz Institute in Rio de Janeiro and the Butantan Institute in São Paulo. Shortly after the Proclamation of the Republic, the National Museum saw its prestige enter a period of steady decline, while other centers began their ascent, offering bigger budgets and additional amenities that could attract the most eminent researchers—a status quo that was not to change until the late 1920s.

The 1920s indeed brought change to Brazil. World War I had ended, as had the optimism of the Belle Époque. The coffee glut and the demise of the Amazon rubber boom in the face of stiff competition from Southeast Asia spelled economic hardship. Anarchist and communist union movements were on the rise, alongside conservative Catholic movements. Modern Art Week, an arts festival held in São Paulo in February 1922, signaled artistic restlessness. Young military officers joined the armed movement known as *tenentismo*, while the Prestes Column engaged in guerrilla warfare and *cangaceiro* bandits ran rampant in Northeast Brazil. In 1922, this turmoil was reined in by a government-imposed state of siege; the press was censored, and the various movements that opposed the oligarchical Republican project were repressed.

From September 1922 through July 1923, the city of Rio de Janeiro was the site of the International Exhibition in Celebration of the Centennial of Independence. Organized by the federal government, which built sumptuous pavilions for the event, the exhibition was intended to convey an image of progress and national union. The government had designed the show in hopes of garnering legitimacy at a difficult time, but by instigating reflections on Brazil's past, present, and future at a moment of serious political crisis, the commemoration in fact seeded unease. Visitors grew more aware of conflict and social tension because the exhibition triggered concern about national construction and about Brazil's place in world civilization. What, after all, was being celebrated? What brand of independence? What kind of nation? What kind of Brazilian people? What type of republic?[10] The exhibition may to some extent have been a paean to the ruling order, but it also awoke society's latent expectations and desire for change. As Hoffenberg has noted, "Exhibitions were meaningful events for participants struggling with the social, political, and economic dilemmas and opportunities of their era."[11]

From the very dawn of the twentieth century, countless intellectuals had criticized the reigning oligarchical regime, holding it accountable for the high-handedness of local authorities and the fact that people had been left to fend for themselves. More voices entered the debate about the roadblocks to nation building. Attention became focused on the vastness of the Brazilian land, on its people trapped in misery, illiteracy, and disease, and on the wholly irrational destruction of its natural riches. The prevailing political and economic liberalism was called into question, decried as excessive, and critiqued for motivating selfishness, while centralization of power was posited as an alternative raised above individual interests. Solutions were proposed for a political and institutional system that demanded more than the mere consensus of the elites and that would transform Brazil's near nomadic population—until then rebuked as inferior—into healthy, educated, and hard-working people, indispensable to the building of a nation. These intellectuals urged society to adopt new attitudes toward nature; Brazilians needed to learn about their country's flora and fauna, its water resources and landscapes—and learn to value them—while the state had to effectively regulate environmental protection areas and national parks and exercise control over the exploration of natural resources throughout the national territory. Based on an authoritarianism characterized by voluntarism and an obsession with education, they believed that if the Brazilian people, in its most genuine expression, could be brought onto the stage through suitable measures, the result would be the emergence of a popular culture duly civilized through learned knowledge and superior reasoning—to wit, "authentic" nationality.

From 1926 to 1935, the National Museum regained momentum under the leadership of Edgard Roquette-Pinto. The institute modeled itself as a prime space for educational intervention and for the coordination of pedagogical projects for the people of Brazil, as well as a place where knowledge was produced. It introduced and enforced a bold and experimental multimedia project. As urban life and consumption became increasingly sophisticated in the city of Rio de Janeiro, where the museum was headquartered, its staff members embraced the era's new means of communication, optimistic that new technologies would allow them to span the chasms yawning between them and Brazil's ordinary men and women, lost in the vastness of their country.

The National Museum was home to a collaborative effort that drew researchers from an array of fields; they engaged in surprisingly varied initiatives that were not confined to the premises of the museum but reached into other institutional and social domains. Staff members like Roquette-Pinto, Alberto Sampaio, and Cândido de Mello Leitão organized public exhibits unprecedented in the history of the institution. They threw themselves into the Biblioteca Pedagógica (Educational Library) editorial project, headed by Fernando Azevedo, and particularly into its Brasiliana Collection, whose ultimate purpose was "to reveal Brazil for Brazilians." They launched the journal *Revista Nacional de Educação*, a forum for science communication aimed at the broader public, whose circulation reached 15,000. They set up a radio station specializing in educational programming and ventured into cinema and the production of educational films. They organized notable events like the First Brazilian Congress for the Protection of Nature, in 1934. They led prolific scientific lives, participating in cultural exchange and attending international congresses. They helped make public policy, including the draft bill for the Game and Fish Code, which lay behind the law decreed by President Getúlio Vargas on January 2, 1934. They joined science associations and other civil society organizations. In fulfilling their "pedagogical mission," the museum staff relied on a range of media, including print, photography, exhibits, movies, and radio programs. Its scientists also maintained close relations with the ruling powers and with other spaces that generated knowledge. Throughout their experiences, these men of science worked and thought collectively, constructing knowledge through frank dialogue. Moreover, they worked to accrue the technical expertise essential to the practical realization of these manifold projects.

The organizational heart of the activities and exhibits at the National Museum was certainly "the Brazilian nation," and the burgeoning of biology as a fully established discipline figured largely in this work. Although the field had existed in Europe since the mid-nineteenth century, it was only in the early twentieth that biology laid down roots in Brazil. The troublesome presence of sick, ignorant, rebellious people was a quantitative and qualitative problem begging for a solution, and biology, as a "master of life," was capable of addressing these ills. It lent itself to a variety of nationalistic practices fashioned within an authoritarian, salvationist political culture.

In the eyes of the museum staff, the field of natural history could describe and name things but could not address the full complexity of life, so it was unable to confront the challenges of Brazilians in distress. Biology, on the other hand, was a decisive form of knowledge, which supported scientific medicine and was based on scientific laboratory practices in the fields of physical anthropology, entomology (especially as applied to agriculture), eugenics, the theories of evolution and genetics, and even phytogeography, zoogeography, and ecology.

At a time when biology was taking shape as a field of its own, separate from (but not better than) natural history, the National Museum endeavored to renew its practices and present itself as an institution in step with the changing world of science. Some of its members also belonged to the Brazilian Academy of Sciences, founded in 1916, which valued specialized experts more than wise generalists, and they worked hard to earn esteem as scientists from specific fields. Yet in its practices, the museum displayed a dynamic and contradictory tendency: although many of its members wanted very much to become specialized scientists, their work with different media formats and with science communication took place in an atmosphere of blurred boundaries between the disciplines.

While striving to make a name for themselves in scientific circles, these scientists also sought government backing for their projects. Most importantly, they wanted themselves and their institute to play an active part in public policy making, and in this way their scientific activities constituted veritable political strategies.

This book explores the budding role of biological knowledge in the construction of Brazilian society from 1926 to 1945, focusing on the National Museum as a strategic institution where a gamut of social actors deployed a distinctive set of practices. It centers on three of the most illustrious among these actors: the arachnologist Cândido de Mello Leitão, the botanist Alberto José de Sampaio, and the anthropologist Edgard Roquette-Pinto. The first year in our timeframe, 1926, sees Roquette-Pinto appointed director of the National Museum. The last year, 1945, coincides with the end of the fifteen-year-long presidency of Getúlio Vargas and with Brazil's return to democracy. It was also the year Mello Leitão, at the apex of his renown as an arachnology expert, made his last trip as a cultural envoy for the Brazilian government on a Latin American mission. That Mello Leitão could win such national and international acclaim lends validity to my proposal to explore the effective specialization of biology as a science and the genesis of activist biology even before biology had been formally recognized as a profession and before there were any special courses for training as a biologist in Brazil.

The investigators at the National Museum identified themselves as biologists well before the field had become established in Brazil, a country where universities came into being rather late. The earliest colleges established in Brazil were the University of Paraná (1912), the University of Rio de Janeiro (1920), and the University of Minas Gerais (1927). Rather than participating in a larger project to inaugurate universities in Brazil, these first institutions merely unified a number of preexisting courses under one formal roof. It was only in the 1930s that broader projects surfaced, like the University of São Paulo (USP), the University of the Federal District (abolished shortly after its creation in 1935), and the University of Brazil, in Rio de Janeiro (1937). The first courses in zoology, botany, and general biology only began at USP in 1934 in the School of Philosophy, Sciences, and Literature, later spreading to the University of Brazil in 1939.[12]

How did the National Museum go about formulating a strategy to reinvigorate its presence on the national political stage? What explains its scientists' success in forging ties with the political powers in place after 1930, within the Provisional Government of Getúlio Vargas? These questions are the guiding thread of the first chapter. My purpose is to understand the overall historical framework and what it was that incited this search for new paths and new practices, with museum scientists as spokespersons. The point is to ask why, and whether, these actions were necessary at that specific moment in history and to explore the demands behind them and the battles in which they were crafted.

It is my contention that men like Roquette-Pinto, Mello Leitão, and Alberto Sampaio were able to represent themselves as scientific authorities in the field and to occupy strategic decision-making positions because the Brazilian state found itself up against the growing challenge of controlling the populations within its territory and because biology came forward as a discipline that could contribute significantly to the solution of political problems. Moreover, these scientists were advocates of tremendously eclectic theories, woven out of eugenics, Mendelism, and neo-Lamarckism; and their anti-Darwinism meshed well with the Vargas government's corporatist political view, which rebuffed class struggle, promoting instead the idea of a harmonious, organic society as the foundation of a new day. Bringing together biology, educational initiatives, and the rejection of social conflict, the National Museum proved successful in its bid to become a respected partner of the government, and particularly of the Ministry of Education and Public Health (MESP) in the early 1930s. Furthermore, the demarcation of biology as a specific field dovetailed with the era's growing nationalism. Biomedicine—where medicine delves into biological investigations and uncovers scientific proof in the laboratory—joined hands with the exercise of biopower. I draw inspiration here from Foucault, who conjectures that matters of life have been of special concern in the exercise of contemporary political power in the West and that power began taking into consideration the biological traits of the human species and formulating strategies for the political management of population phenomena like birth, reproduction, disease, and death. In Brazil, the act of governing came to include the life problems faced by people scattered across a vast national territory, while biology firmly established itself as a source of indispensable knowledge.[13]

The quest to understand the circumstances under which the National Museum and its scientists developed their practices uncovers a political skirmish over which social rules and which institutions would hold sway. If the National Museum was going to portray itself as an institution that should have a say in deciding the direction of the nation, its scientists would have to negotiate around the many other political projects on the agenda. Numerous victories ensued but so too did crushing defeats, evincing the historical complexity of those years. In some cases—such as the eventual enactment of the Game and Fish Code—the museum's scientists ended up having much less influence over government decisions than they would have liked.

The second chapter explains how some members of the National Museum set their sights on entering the game of politics and realizing their dream of actively assisting in the education of the Brazilian people. The Vargas Provisional Government (1930–34) warmly welcomed the work being done at the National Museum. At the museum, the new government encountered a sophisticated framework of science communication and education, organized through the museum's Assistance Service for the Teaching of Natural History, a department created by Director Roquette-Pinto at the outset of his mandate, in 1926. One of a kind in its emphasis on popular education, the National Museum presented itself as a strategic setting for educational action. It proved to be in tune with the government's desire to mold a new "Brazilian man" and received broad government support through the MESP.

The use of new technologies and the diversity of media were central features of the museum's activities during those years. As mentioned earlier, the museum participated in radio programs; organized exhibits, courses, and workshops; utilized slides for educational purposes; produced and screened educational films; and published magazines and books. Treating patrons to interactive experiences on its own grounds, the museum also breached its institutional walls and went directly inside public classrooms. Through these initiatives, and by devising new practices for producing and disseminating knowledge, the National Museum negated the prevailing notions of what natural history museums should be.

During the period bracketed by these halcyon days and by the mounting challenges faced when the Vargas administration altered its educational policies, the museum played an ambiguous role. Its members moved easily between natural history and biology and between scientific specialization and knowledge built at the intersection of different fields, through hands-on interaction with technical and artistic techniques. While these practices quite likely account for the rich tapestry of accomplishments at the National Museum in those years, these same factors were the chink in its armor when Gustavo Capanema, as the new minister of education and public health, launched an educational reform and spurned the museum as a locus for the production of knowledge. When the government headed in these new directions, the scientists at the museum fell out of its favor, and the work being done inside the museum's walls lost steam. Enthusiasts like Roquette-Pinto and Mello Leitão had to relinquish their posts and find new routes for carrying their activities forward.

The third chapter illustrates how biological knowledge and practices played a weighty role in the game of politics during this period. Brazil offered no biology courses at that time, and biology was not even a formal profession. It is not my intent to define the precise moment it did emerge as a field. Rather, I want to show how biologists were formed where webs of relationships converged during a time of evolving practice. As a man of his time, Mello Leitão offers us a perfect example of the making of a scientist specializing in biology during those years. Yet he was so unique that he also stands as an exception, a rare gem. The realization of these two facets of the figure of Mello Leitão leads to an important observation: when we study history, we discover that social actors, in the process of confronting diverse interests, defining strategies, and engaging in dialogue with specific historical conditions, surpass existing determinations. As they tackle their contemporary challenges in a period of social struggle, they produce something new and original. In other words, the writing of history gives us the chance to reflect on the creation and transformation of human societies.

Mello Leitão, Roquette-Pinto, and Alberto Sampaio were amazingly creative in their professional lives. We will find them busily involved in political projects and in countless civil and scientific associations of weight, while they deftly carved out an innovative niche for the National Museum. They were attentive to the question of nature, receptive to new technologies and means of communication, and open to different fields—even as they yearned for specialization and invested in it. And yet, by idealizing the Brazilian people and hoping to mold them to their own expectations, they reinforced authoritarian perspectives.

It is tempting to see these scientists as the pioneers behind many of our contemporary practices. It would be no stretch to paint them as trailblazers in a wide variety of arenas: environmental education, the renewal of museums through connections with universities and extension work, science communication and the establishment of public science projects, distance education, transdisciplinarity, the creation of university radio and television networks, the struggle for social inclusion through education, and advocacy of the sustainable exploration of nature. But this would mean ignoring or glossing over the authoritarianism inherent in their practices and possibly being induced to slip into the same logic. And we would be eschewing Marc Bloch's invaluable insights about the need to scorn myths of origin, examine our differences in relation to the past, and see the potential for inaugurating new practices.[14]

It is precisely from the perspective of difference that the history of the National Museum during these years should be researched and studied. If we weigh what distances us from the actions of these scientists rather than simply singing their praises, we improve our own chances of acting creatively in the present and of making a real break with the authoritarian tradition they reaffirmed—their incredible talents for innovation notwithstanding. Writing history can be an exercise in political praxis while also being motivated by a desire for social transformation.

And so it was that the idea for this book was born, when I, both moved and stunned, waved back at the destitute children who dwelled alongside the breathtaking tourist attraction called the Lençóis Maranhenses.

1

ACTIVIST BIOLOGY

Today we breathe easier. The laboratory has given us the argument we so eagerly sought. Grounded in it, we shall counter Le Bon's sociological condemnation with the higher voice of biology.

—MONTEIRO LOBATO, *PROBLEMA VITAL*, 1918

A REPUBLIC OF SCIENTISTS

IN FEBRUARY 1933, the front page of Rio de Janeiro's newspaper *A Noite* featured an interview with the zoologist Cândido de Mello Leitão (1886–1948), one of the authors of a bill to regulate hunting and fishing in Brazil. The Ministry of Education and Public Health (MESP) had assigned this task to a committee comprising Mello Leitão and two other members of the National Museum: the botanist Alberto José de Sampaio (1881–1946) and the anthropologist Edgard Roquette-Pinto (1884–1954), the museum's director.[1]

As Mello Leitão explained in his interview, "all cultured peoples" had made the defense of fauna and other natural riches a priority, and Brazil would be taking the same path by following through on this government initiative. Of course, the zoologist admitted, the proposed measures would displease those who "exterminated birds" and the merchants who put their "own, limited interests ahead of the name and interests of Brazil." But it was a time of "patriotic reconstruction," as he saw it, and everyone had to think first of his country and only second of himself. While the law could not completely disregard economic interests, its chief concern had to lie with protecting fauna—"the nation's heritage" and the "charm of its landscape."

The 1920s had been years of major upheaval and great expectations in Brazil, with its oligarchic republic suffering serious blows. Coffee prices fell, opposition groups formed among both the military and the urban middle classes, and there were pockets of more militant social unrest. At the close of the decade, a political and military movement that proclaimed itself the Revolution of 1930 swept Getúlio Vargas into power. The Vargas Provisional Government was quick to promise deep change, even while grappling with the ideological differences that arose among its backers. The regime fed hopes for national construction and for the settlement and true conquest of Brazil in all its immensity; it vowed to work for the benefit of those living in urban and rural Brazil alike, and to afford those in the farthest reaches of the land a healthy life of labor and civic participation—all orchestrated by a strong, central, organizing state. The new government won the support of intellectual groups long critical of the oligarchic republic and its failure to cure the woes of the ailing, illiterate people spread across Brazil's untrodden *sertões* (hinterlands), strangers to the lavish life of Rio de Janeiro, then the nation's capital. Furthermore, the republic had failed to steer urban workers away from the dangers of communism.[2] The state set its sights on organizing Brazil's citizens, its land, and everything it held: people, landscapes, soil, water, flora, and fauna.

This was the "patriotic" backdrop to which Mello Leitão was alluding in the newspaper interview when he highlighted two key benefits of the bill submitted by the National Museum committee. In the first place, the bill itself would stimulate the production of scientific and zoogeographic knowledge, which could be used to guide future improvements to the law. In the second place, the bill valorized Brazil's national scientific institutions by assigning them a primary role in the collection of the country's fauna treasures and by controlling the activities of foreign scientists. After all, Brazilian researchers were tired of having to travel abroad to study their own country's fauna, gathered by foreign naturalists and then unceremoniously hauled off to Europe's great museums.[3] The interviewee also advocated the urgent adoption of other measures in addition to hunting regulations, in step with the example set by other countries where fauna was protected in nature parks.

The MESP had asked the National Museum to draw up the bill in August 1932. Roquette-Pinto submitted the draft two months later, prefaced by a twenty-four-page document laying out the presuppositions of the three scientists. In their prologue, they argued that hunting laws were indispensable and that Brazil was among the most backward of the major countries in this area. They warned that merely handing down decrees would not suffice; implementation had to be guaranteed on two fronts: first, the General Directorate of

Public Instruction had to make the protection of nature a top priority; second, enforcement had to be stringent, given the mammoth size of Brazil and the short-term impossibility of "taking the benefits of protective laws into the far-off sertões." For the drafters, the most efficacious, modern example was in Italy, where Mussolini had instituted the Italian Forestry Militia, a military-like unit commanded by a general and under the direct supervision of Il Duce.[4]

According to the bill, because hunting provided an important food source for inhabitants of the sertões, it was connected to the issue of health and therefore should not be banned outright but only regulated. Regulations should stipulate hunting seasons in accordance with each species's breeding cycle but should also take into account which animals might be considered beneficial resources and which were nuisances. Regionally, certain species would also have to be monitored for overpopulation. All this would require scientists to furnish ongoing input as consultants and participate directly in decision making, since "research on the protection of nature is an educational and scientific specialty."[5]

The bill was published in Brazil's *Federal Register (Diário Oficial)* one month later, in November 1932, without the preamble. The general public was invited to offer suggestions to the MESP.[6] The document assembled by the three scientists was rather bold, in that it ran counter to economic interests and claimed a decisive role for the National Museum and its members. The commission gave as its paramount justification for regulating hunting the fact that Brazil's wild fauna was thinning out and its national forests needed protecting. Under the new law, all animals and birds within Brazilian territory would be the property of the state, and there would be a nationwide ban on hunting animals that were deemed beneficial resources. The National Museum would draw up a list of the latter, along with a list of nuisance animals whose killing would be allowed under the law. Hunting would be banned inside urban and suburban areas whose populations numbered more than 15,000. All hunters would have to carry licenses. Municipalities would issue seasonal licenses for sport hunters, while the National Museum would be responsible for issuing scientific licenses. A special license, valid for a limited time only, would be required to stalk big cats.[7]

Hunting seasons would be stipulated according to the preservation needs of the various species. The sale of weapons, certain ammunition, and hunting dogs would be banned outside of these seasons, as would the sale of wild animal pelts, hides, feathers, or horns along roads and railways. There would be an immediate five-year ban on the hunting of endangered species, such as otters, egrets, rheas, capybaras, seriemas, and deer, and no products from these animals could be sold, transported, or exported for the duration of the ban. The capture of diurnal butterflies in the Federal District would also be prohibited for five years. Lastly, the government was to sign treaties with bordering nations to help foil the illegal export of products from any animals protected under the law.

The bill provided for a number of educational initiatives tailored to the public at large. Ecological stations were to be set up at national parks so that the ecology and etiology of wild animals could be studied. From elementary through high school, curricula were to include studies on wildlife and its protection, while radio stations would be mandated to broadcast a weekly program on the conservation of Brazilian fauna, especially birds and mammals with resource potential. Minors were also expressly banned from taking part in hunting.

As audacious as the bill was, it was not without precedent in Brazil. From the dawn of the republican period, a number of scholars who were outraged about the slaughter of certain species had demanded that authorities do something to protect the country's fauna. In 1895, writing as director of the Pará Museum of Natural History and Ethnography—a post he held from 1894 to 1907—Emílio Goeldi filed two complaints with the governor of Pará to protest the "barbarian annihilation of egrets and scarlet ibis." These birds—described by Goeldi as "captivating adornments of nature from the majestic Amazon River"—were being decimated by hunters who would ambush them in their breeding grounds, such as Marajó Island, "in order to rip from them their few feathers, thousands of which are needed to make one kilogram of this heinous merchandise." Goeldi asked for a ban on the hunting of egrets and scarlet ibis from June through January, the protection of nests on public land, and the levying of hefty taxes on the sale of feathers. He also issued a special plea for the government of Pará to pressure Congress to approve hunting regulations.[8]

Hermann Von Ihering, director of the Paulista Museum from 1894 to 1915, had been another proponent of a federal law against predatory hunting. He was in favor of maintaining a list of animals that could only be hunted during specific months of the year. He also backed an outright ban on the hunting of birds that were deemed of potential benefit, a ban on the sale of feathers, pelts, and hides, and a minimum hunting age of eighteen, since he felt that hunting could jeopardize the character formation of minors.[9]

The discourse of these two foreign-born scientists—both naturalized Brazilians—also encompassed the idea of nature as a national asset and the need for a kind of "pedagogy of national nature."[10] In the early days of the republic, these arguments carried political overtones, because they called attention to how these hunting practices were affecting the Brazilian nation and criticized the era's prevailing liberal models. In Goeldi's opinion, these were "shocking crimes, committed against the nature of this beautiful Country"; the vile feather trade brought about the moral downfall of its accomplices, lost to barbarity and guilty of "heinous murder" and the looting of "the sacred, intangible heritage of our Nation." Von Ihering condemned the orthodox liberalism of the federalist republic and called for stringent federal laws to defend the "true interests of the Nation" and for the creation of biological reserves, stations, and parks, following the examples set by other countries.[11]

The bill as finalized by Roquette-Pinto, Mello Leitão, and Sampaio in 1933 thus incorporated themes and ideas already on the scholarly agenda. But it was completely unprecedented in one regard, which had to do with the context of Brazil's post-1930 republic and the process of building a strong central state. Goeldi and Von Ihering had written in scientific journals or had addressed letters to government authorities expressing their indignation over what they saw happening. But the text by the three scientists from the National Museum was published in the *Federal Register* as an MESP project, open to suggestions from the public—in other words, it was an official document that had originated from a government agency. So while there was nothing extraordinary about what the three of them said as scientists, it was quite an achievement for them to be saying it from this platform.

There is no doubt that hunting remained unrestrained in post-1930 Brazil, as confirmed by a glance at some of the era's most popular magazines and newspapers. Stories of the hunt were recounted in a tone of dramatic adventure in *Noite Ilustrada*, a supplement of the Rio de Janeiro newspaper *A Noite*. News reports covered the busy, profitable trade in hides, leather, and horns. In 1931, for example, the export firm Casa Mastwyk sold a single shipment of lizard skins, jaguar hides, and hides of other wild animals to Europe for 380 contos de réis, an astonishing sum.[12] In the accompanying photograph, a group of men seated on large bales readied for shipment abroad show off a sizable number of hides and pelts (figure 1). Other news stories extolled the virtues of hunting as a sport that demanded and developed character traits distinctive of a certain type of masculinity, like courage, skill, and cold-bloodedness. There was much fanfare about the 1931 visit to Brazil of Alexandre Siemel, a veteran European jaguar hunter who ventured into the hinterlands of Mato Grosso on what promised to be his greatest adventure. The young man was photographed next to the trophies of his bravery: large slain jaguars (figure 2). In Mato Grosso, planters were famed

FIGURE 1. Shipment of hides and skins for export. *A Noite Ilustrada*, May 20, 1931. Courtesy of Hemeroteca da Biblioteca Pública Luiz de Bessa.

for their hunting feats, like Jacques Ribeiro da Luz, who had killed thirty-nine jaguars in less than two years and had them stuffed to decorate his manor in Santa Maria.[13]

Signs of the hunt were visible in the most sophisticated settings of Brazil's cities too. Jackets and hats fashioned from animal pelts, adorned with feathers or stuffed birds, and wallets and belts of exotic leather were coveted by the urban middle classes, who wore them to proclaim that they belonged to the world of chic.[14] By associating the sport with a life of nobility and luxury, the appeal of the hunt was directed at city dwellers, far from the backwoods. During the decades that separated Goeldi's and Von Ihering's protests from the bill drafted by the National Museum scientists, the representatives of major manufacturers of hunting weapons, like Winchester, Standard, and Hunt, had entered the market, and greater attention was paid to hunting for sport. Advertisements featured photographs and prints of hunters decked out in their gear, posed against peaceful countryside backdrops in the company of handsome hunting dogs (figure 3). Hunt clubs organized hunting parties on planters' lands. Hunting manuals, filled with descriptions of remarkable exploits through the forests of Brazil, proffered detailed instructions on trapshooting and the use of guns and ammunition. This is probably one reason why the definitive version of the bill

— 2 — A NOITE (Supplemento)

NA SELVA DE MATTO GROSSO
ALEXANDRE SIEMEL
MATADOR DE ONÇAS

A caça ás féras é um desporto arriscado, e que, por isso mesmo, não conta grande numero de adeptos. Não exige apenas bravura pessoal, mas um conjunto de aptidões, como a pertinacia, o sangue frio, a resistencia physica, habilidade e boa pontaria. Exige, ainda mais, a intuição especial sem a qual não é possivel um caçador realmente de carreira. Só o caçador de temperamento, possuindo a paixão da caçada, póde realisar partidas notaveis no genero e constituir sobre a actividade cynegetica uma historia. E só elle, de outro lado, será capaz de padecer os precalços da caça sem esmorecimento. No mesmo campo, varios individuos egualmente destemerosos fracassarão se a nenhum assistir o instincto do caçador — essa acuidade transcendente que leva o individuo a realisar no terreno os passos que lhe propinarão boa e segura presa. O que caça pelo méro proposito de caçar, esse é um fracassado e não raro cae no ridiculo.

* * *

No numero dos caçadores de coração está o Dr. Alexandre Siemel, que ultimamente passou pelo nosso porto como membro de uma expedição de caça e de estudos que operará na selva matto-grossense. Antigo explorador do continente sul-americano, esse matador de onças é completo no genero. Além das capacidades naturaes, Siemel conta com uma experiencia longa de caça ás féras. Segundo declarou, conhece o manejo de todas as armas rudimentares do indigena americano, como o arco, a flexa, a lança ou "tarope", gabando-se de ser o unico europeu que haja abatido féras utilisando-se da lança.

Siemel, que seguiu para o "hinterland" mattogrossense, conta realisar agora a melhor batida da sua

ATE' AS CREANÇAS...

Os pequeninos estavam afflictos para vêr o papae vestir a roupa que veio da

ALFAIATARIA GUANABARA

54 - Rua da Carioca - 54
Rio de Janeiro

PARA TODOS OS FINS DE ILLUMINAÇÃO

FIGURE 2. "In the jungle of Mato Grosso, Alexandre Siemel, jaguar slayer." *A Noite Ilustrada*, January 28, 1931. Courtesy of Hemeroteca da Biblioteca Pública Luiz de Bessa.

felt it had to discourage gentlemen hunters by banning hunting in larger urban centers; this same interdiction was meant to curb the actions of the poor who tried to fatten their larders by hunting small animals or birds.[15]

The image of the hunt as an exciting adventure was directed toward a diverse audience. In 1933—the year the final touches were put on the hunting bill—the writer José Bento Monteiro Lobato published a popular children's book entitled

FIGURE 3. Ad for hunting rifles. *Fon-Fon!* 3:33 (1909): 6. Courtesy of Hemeroteca da Biblioteca Pública Luiz de Bessa.

Caçadas de Pedrinho (Little Peter's hunting adventures), part of a multivolume series devoured by generations of young readers. Lobato was a highly influential writer, journalist, and publisher and had an active voice in the day's political and social debates, but his greatest and most enduring success was his literature for children and young adults. In *Caçadas*, the biggest, mightiest jaguar from the woods near Yellow Woodpecker Farm succumbs to a sequence of assaults by memorable characters: the Viscount of Sabugosa strikes him with a buccaneer's sword; the ragdoll Emília uses a shish kebab skewer as her weapon; young Narizinho attacks him with a knife; Narizinho's pal Pedrinho thunks him with the butt of his shotgun (it had misfired first time around); and the piglet Rabicó shoots him point-blank. At the end of the battle, the heroes celebrate the mighty cat's death to the sound of hurrahs and war songs. Those were indeed other times, when we had very different feelings about animals.[16]

The proposal to effectively control hunting would step on the toes of private interests, as Mello Leitão had pointed out in his interview. In addition to the countless hunters across the sertões and woodlands of Brazil, there was a complex web of buyers and intermediaries involved in trading pelts and hides, mostly for illegal export. But the measures would also have a marked impact on buying habits that were fast becoming ingrained among the most privileged elites and even among the rising middle classes, avaricious for status symbols. Hunting regulations would affect the expanding trade in imported weapons and ammunition in Brazil. Furthermore, they would likewise affect the world of fashion, where upscale stores did a brisk business in fur coats and stoles and in hats embellished with animal parts (figure 4). Other industries, like manufacturers of horn combs, depended on the byproducts of slaughtered wild animals. In 1930, there were some 3,500 Brazilian businesses making leather goods such as belts, wallets, handbags, bridles, suitcases, and steamer trunks (consumption of the latter climbed as tourism to Europe rose). Hides, pelts, and leather ranked third on the list of Brazilian exports from 1898 through 1918 and second from 1919 through 1939.[17]

More to our point, the bill addressed the nation's interests and its natural heritage. It assigned the National Museum a place that raised its scientists to the status of agents qualified to watch over this public asset, side by side with the state. The bill also reflected the dream of a "republic of scientists," where knowledge of nature would be seen as indispensable to the Nation. But the question becomes: what made it possible for these scientists not just to offer a systematization of protection measures—as Goeldi had done in 1895 and Von

FIGURE 4. Ad for a women's clothing store. *Fon-Fon!* 13:30 (1919): 9. Courtesy of Hemeroteca da Biblioteca Pública Luiz de Bessa.

Ihering, in 1902—but to work inside rather than outside the framework of institutional political power? What compelled these men to wager so heavily on a bill that clashed with conventional habits and sentiments, and with entrenched economic interests? What gave them enough confidence to believe they might upend Brazil's destructive and shortsighted exploitation of its environment and mandate new parameters, in consonance with the post-1930 administration's agenda of "patriotic reconstruction," as Mello Leitão put it? Biological knowledge had begun to attain political status in Brazilian society in the early twentieth century, thanks in part to the work of scholars like Goeldi and Von Ihering. I will argue that from the time these two men lodged their demands to when National Museum scientists began working alongside the Vargas Provisional Government, biology had achieved firm footing in Brazil as a specific field of knowledge. It had earned its place as a "master of life," qualified to provide the country with decisive knowledge, and had come to play a truly notable role in Brazilian society, owing to factors both at home and abroad.

This chapter begins by exploring the emergence of biology as a field distinct from natural history, starting in the early years of the twentieth century when a series of circumstances fostered the valorization of biological knowledge and the growth of institutions dedicated to its study in Brazil. Next it examines how scientists submitted their demands to the government while also defending the centralization of power in the hands of a state that appreciated their knowledge as a public policy resource. Because the National Museum played an important role in this process, our focus will be on this institution and three of its most eminent scientists with ties to biology—the same three who were on the committee that drew up the hunting and fishing bill. All three were of authoritarian bent, with similarities in their intellectual backgrounds as well, particularly the fact that they were staunch anti-Darwinists. Their fierce rejection of the notion of the struggle between species and their vision of a harmonious, organic society formed the basis for their deep affinity with the Vargas regime and the path it was taking—a clear example of how scientific reason has its own history.

THE AGE OF BIOLOGY

The term "biology" was introduced around 1800 by three scholars working independently of each other, the most famous of whom was the Frenchman Jean-Baptiste Lamarck. Yet as Ernst Mayr has pointed out, biology, in the contemporary sense, simply did not exist as a field of knowledge at that time.

Rather, it evolved over the course of the nineteenth century as its main areas were systematized: embryology, cytology, physiology, evolution, and genetics.[18]

Michel Foucault argues that we can write no history of biology prior to the nineteenth century because it simply did not exist until then. He cautions against any history of science that detects supposed linear paths between forms of knowledge, chaining them together in a kind of progressive unfolding of truth. He instead proposes a genealogy that detects discontinuities and highlights turning points and the disintegration of lines of thought. The French philosopher believes that biology, as the science of life, differs radically from natural history. In his mind, the latter came into being fused with the meticulous morphological description and classification of living organisms; as a branch of knowledge dedicated to naming what was visible, natural history intermingled knowledge about plants, animals, and minerals. It surfaced out of accounts by travelers who journeyed long distances and trekked through unknown lands to then return to their museums and botanical gardens, their trunks packed with faded exsiccatae and animals that had been stuffed or preserved in glass jars, along with samples of rocks, stones, skulls, and indigenous artifacts, all carefully transported across oceans.

Other epistemological conditions would give birth to biological knowledge. Foucault identifies these conditions in the comparative anatomy of George Cuvier, who "was to topple the glass jars" at the National Museum of Natural History in Paris so that he could dissect "all the forms of animal visibility that the Classical age had preserved in them," investigating the hidden function of organs and their anatomy with a special emphasis on the organism and its inner logic.[19] This set the stage for Charles Darwin's breakthrough. Systematized some decades later, his theories revolutionized our understanding of life by positing the study of populations as a unique source of insight into the bonds between environment and organism.[20] Biology thus appeared as a field broader than natural history, even though the latter was often described as the "origin" of the younger science or was even viewed as a subfield. Biology began coming into its own in the mid-nineteenth century and continued its journey throughout the twentieth as part of a process where scientific practices became interwoven with complex social relations.

One of Foucault's main goals was to criticize the linear view of the history of science, which sees a continuous progression from natural history to biology, and he therefore laid great emphasis on the disjuncture between the two fields. Yet he erred when he disregarded the dynamic nature of natural history, the potential for natural historians to act not as amateurs but as scientists, and the field's relevance to biology not only through the twentieth century but even today.

Some authors have questioned the division between a natural history written by amateurs, antiquated and destined to fade away, and a "scientific" biology, developed by professionals, experimental and technological in nature. As these scholars have it, there is greater continuity between the two fields than meets the eye. Biology could be metaphorically represented as a landscape "in which territory is contested, divided, reunited, and its boundaries redrawn over time by competing (and sometimes co-operating) groups."[21] Natural history would thus continue to stand as one of the distinct elements within this landscape.

Although we cannot claim a linear scientific progression from natural history to biology, this viewpoint nevertheless predominated among the scientists who emerged as the "new" biologists in the early twentieth century; further, it was the narrative underpinning the quest for a new status on the part of scientists devoted to the study of life. This was true as well in Brazil, where touting biology as a field superior to and distinct from natural history was common practice among scientists who wished to valorize their own knowledge and distance themselves from the image of "mad" naturalists or mere amateur collectors of natural objects.

In 1918, Roquette-Pinto observed that in Brazil the study of species was growing more specialized and acquiring a physiological outlook; scholars were no longer content with just "describing the forms and groupings . . . of plants and animals" but were striving to learn "the secrets of the dynamic existence of each."[22] And while biology was thus developing into a distinct discipline, its relations with political and cultural life were no less important than the framing of its theories, supporting the idea that the history of a science is necessarily a history of the society in which this science comes to life.

The persistent effort to distinguish biology from natural history was part and parcel of the endeavor to make the world at large value the biological sciences. Researchers like Roquette-Pinto at natural history museums worked to dispel the notion that they were mere collectors of exotic things. The idea of rupture with past knowledge was often emphasized as a strategy for persuading the world that the knowledge produced at museums was entirely new and compatible with the paradigm of scientific research conducted in laboratories. Yet in practice, morphology and classification remained the fundamental subareas of botany and zoology, while natural history delved into evolution, paleontology, ecology, and biogeography. Only much later would biologists realize the importance of this overlap. In the first decades of the twentieth century, naturalists anxious to be identified as scientists rather than as collectors of taxidermied animals or preparers of dried plant specimens sang the praises of biology—tantamount to a rejection of natural history.[23]

If natural history helped prop up the great modern empires,[24] biology lent itself well to nationalist longings, starting at the end of the nineteenth century. In large measure, relations between biology and society revolved around nation building. At the outset of the twentieth century, an intellectual movement in the United States unified biologists, educators, media moguls, political leaders, and publishers in support of the role biology should play in building modern American life, conjoining scientific understanding, philosophical convictions, and a stalwart nationalist outlook. Among the most influential of these thinkers was John Dewey, an advocate of a pragmatic brand of education. Biology was heralded as providing a strong basis for the nation's foundation, one that could help develop the intelligence and skills of U.S. youth and produce aware citizens. U.S. scientists and intellectuals criticized classic education and deemed it incapable of coping with the complex needs of modern-day life or with future challenges. They championed Dewey's pragmatic education, grounded in life itself, and believed few fields could serve the cause as proficiently as biology. It was the "age of biology," binding together life, nature, land, and the formation of future citizens.[25] Leading researchers in the United States reached out beyond the ivory tower to join with high school teachers in science education initiatives, determined to make biology a vital field in the daily lives of America's rising urban middle classes. They devoted themselves to schools for the young, founded educational magazines, enlisted in local, state, and national committees to value biology in education, published textbooks, debated pedagogical practices, and redesigned the physical spaces inside natural history museums, turning them into sites for experimental teaching.

By the 1920s, much of mainstream America had become familiar with notions of biology. The movement's greatest expectations were that biology, as "master of life," would supplant superstition and folk wisdom; demonstrate the unity and interdependence of animal, plant, and human life; and equip citizens to adapt to a vast array of circumstances and transform their environments. Biology presented itself as the "promise of American life." The "gospel of biology" offered sure tenets for guiding human behavior, based on the belief that nature was normative. The biologist came on the scene as a kind of "prophet" or even "healer," who revealed nature's moral teachings and its political and ethical lessons to the rest of society. Teaching and research institutions in New York, like DeWitt Clinton High School and the American Museum of Natural History, lent new impetus to biology.[26] Anísio Teixeira, the first translator of John Dewey in Brazil and pioneer of the Escola Nova movement, studied at the Teachers College at Columbia University from 1928 to 1929.[27]

In Latin America, the formation of biology as a specific field was also linked to larger historical and social processes, particularly to nation building in liberal states in the early decades of the twentieth century. Colombia, Venezuela, and countries in Central America and the Caribbean were enjoying fast and hardy development thanks to commodity exports, like coffee, bananas, sugar, tobacco, and rubber. Public authorities celebrated the wealth and excitement engendered by this scenario, while local elites demanded that the government safeguard the continued profitability of the agricultural and extractive industries, threatened by frequent environmental catastrophes.[28]

For Brazil's biomes, these years witnessed a formidable increase in farmable land and the concomitant destruction of expansive swathes of forests to make way for plantations. Experimentation with seedlings and seeds also increased, as did their circulation, and so diseases caused by insects and fungi readily erupted into epidemics, often times wreaking not only economic damage but social and political damage as well. Both planters and public officials, in their recurrent attempts to defeat the serious crises triggered by the pathogen-caused destruction of agricultural landscapes, often consulted with scientists experienced in botany, entomology, epidemiology, and ecology. Although individual nations faced these repeated epidemics alone, there was a transnational component to the spread of disease and pests and to the search for solutions. During times of crisis, top U.S. research centers became major reference points for a number of Latin American nations, spurring intense exchange of biological knowledge and helping boost appreciation for this field of science.[29]

In Brazil, biology continued its development as a new field, with its hands in an increasing number of projects for the benefit of society. However, all this transpired under highly contradictory conditions, of which historical research affords us only a fragmented and not always coherent picture, testifying to the complexity of the history of Brazilian society during this time.

EUGENICS AND MISCEGENATION

Biology's emergence in Brazil was entwined with the reception of eugenics, which began winning favor in Brazilian intellectual circles during the final decades of the nineteenth century. There were heated debates over the inferiority of the country's mixed-race population and the possible solutions to this so-called problem. Mendel's theory was repeatedly invoked to justify racist determinism but likewise to argue in favor of the genetic advantages of miscegenation. A major point in the spread of these ideas was the First Brazilian Eugenics Congress, held in Rio de Janeiro in July 1929. Roquette-Pinto, a noted participant at the event, stated that eugenics always fueled fiery discussions because it stood at the intersection of biology and "social, political, religious, and philosophical issues and . . . prejudice." His criticism of his contemporaries' determinist stance notwithstanding, he recognized that, "for modern nations," no issue was more important than that of population, for everything depended "*on people*, on their number and quality" (emphasis in original).[30]

Francis Galton, the father of eugenics, was not a true biologist but rather a polymath who worked in a number of fields and made major contributions to statistics, meteorology, and geography. A cousin of Charles Darwin's, he was active in the heated debates over evolution that followed publication of *On the Origin of Species* (1859). Based on a loose interpretation of the book's first chapter—where Darwin looks closely at the practices of breeders of domestic species who try to develop predetermined characteristics—Galton organized ideas on the hereditary character of human abilities and applied statistical methods to individuals and, most importantly, to populations. Yet he cannot be called a disciple of Darwin, because he espoused a number of contrary views, such as the concept of evolution by jumps, along with formalist and essentialist notions. He argued that the quantitative measurement of variability and the definition of statistical laws could be applied to a wide range of social issues. He also rejected the idea that an individual's moral and educational formation could supersede or even temper his or her innate traits. Galton believed that by confining and sterilizing some people and stimulating others to reproduce, it might be possible to shape populations that boasted superior hereditary and biological traits.[31]

The penetration of Galton's work into Brazil added new fodder to the racial debate in that country, where the question of the inferiority of some populations had long been discussed. The topic can even be viewed within the broader framework of the "New World polemic," dating back to the ideas of naturalists like the Comte de Buffon, Johann F. Blumenbach, and Louis Agassiz.[32] In the late nineteenth and early twentieth centuries, the urgent need to respond to serious labor problems revived the racial tone of the discussions; voices were often heard defending "whitening" and population control as ways of perfecting

Brazilians—to use the language of the day. In 1888, in the midst of impassioned debates about replacing African slaves with white laborers, Brazil became the last country in the Americas to abolish slavery. There was profound pessimism about the potential of the black and mixed-raced populations, who were seen as degenerate, impossible to civilize, and condemned to decadence. Deeply rooted social practices became the target of criticism, like reliance of well-to-do white women on wet nurses, who were generally black. As physicians resolutely combated the practice, maternal care became a matter of public debate in which mother's milk assumed a symbolic dimension.

In this context in the 1910s and 1920s, eugenics won many followers among Brazilian intellectuals. One of its most radical proponents was the physician Renato Kehl, whose controversial suggestions included amending the civil code to control marriages between people deemed "biologically imperfect." Mello Leitão was another zealous disciple of eugenics in the early 1920s. He expressed his vexation over how hard it would be to achieve this "splendid utopia," whose greatest roadblocks were love, ambition, and democratic regimes. The necessary reforms were unpopular, he said, and government leaders who depended on the vote would be unable to implement them.[33]

Other views, however, came along to counter the eugenicist argument. In 1918, Monteiro Lobato wrote that he was relieved to say that the "higher voice of biology" had saved Brazilians from the obstacles inherent in the notion that miscegenation condemned a society. Curiously enough, he himself had contributed to this fatalistic vision of the Brazilian population a few years earlier, in 1914, when he wrote the short story "Velha Praga" (Old pest), published in *O Estado de S. Paulo*, then one of Brazil's most influential newspapers. The tale drew a detailed portrait of the Brazilian *mestiço* (person of mixed descent) as a "woodlouse," personified by the character of Jeca Tatu, a backward man, physically and morally incapacitated, for whom there was no hope and no future.[34] Monteiro Lobato was soon to reverse his stance, however, and recant his earlier argument that *mestiços* were inferior by nature. In 1918, also writing in *O Estado de S. Paulo*, Monteiro Lobato declared that—loosely translated—"Jeca is not the way he is by nature but by nurture" (Jeca não *é* assim, ele *está* assim). Flying in the face of the pessimistic predictions of the apostles of Gustave Le Bon, Monteiro Lobato wrote that all the Jecas in Brazil could successfully transform themselves from poor, sickly denizens of the interior into prosperous planters, given the right conditions, the practical application of the results of medical experimentation, and sweeping government initiatives in basic sanitation. The lack of health care and rampant illiteracy became the standard explanations for the physical and mental weaknesses that the day's intellectuals and doctors attributed to Brazilian populations, particularly those in rural areas. The microscope promised the redemption of Jeca Tatu.[35]

MICROBIOLOGY AND PUBLIC HEALTH

The steadily growing importance of biology in Brazilian society also owed much to microbiology's success in solving public health and hygiene problems and to medicine's adoption of the biological method of experimentation.

The successes of microbiology and the recognition of Pasteur's discoveries by the scientific community brought a late nineteenth-century renewal of Western medical practices. An outgrowth of biology's experimental methods, scientific medicine was grounded in laboratory research and formed the basis for studies in immunology, biochemistry, parasitology, and bacteriology, to name a few; biology had anointed the living organism as an object, "be it a superior organism or a microorganism, man or cell."[36] It presented itself as a strategic ally to public authorities grappling with the serious problems caused by population growth, urban crowding, and new environmental conditions, which bred epidemics and their chaotic consequences. Alongside the hospital, the laboratory was now a prime setting in medical practice.

In Brazil, the first medical schools appeared in the nineteenth century: in Salvador in 1808, Rio de Janeiro in 1808, and Porto Alegre in 1898. At the dawn of the twentieth century, schools of medicine were founded in Belo Horizonte (1911), Curitiba (1913), Recife (1915), and Belém (1919). Microbiology had been a specialized field in Brazil since the twilight years of the empire. In 1888, Dom Pedro II inaugurated a Pasteur Institute, attached to Santa Casa de Misericórdia in Rio de Janeiro; the only work done there, however, was to replicate the French rabies vaccine. The Bacteriological Laboratory opened its doors in Rio in 1892. Headed by Domingos Freire, professor of biology and organic chemistry at the Rio de Janeiro School of Medicine, the purpose of the lab was to conduct bacteriological research. It was there that the scientist Oswaldo Cruz picked up the basics of microbiology as an assistant specimen technician (*ajudante de preparador*).[37]

In São Paulo, the Vaccinogen Institute and the Bacteriological Institute were established in the early days of the republic, in 1892 and 1893, respectively.

The former made vaccine while the latter, headed by Adolpho Lutz, advanced the fight against diseases such as yellow fever, typhoid fever, the bubonic plague, cholera, and smallpox. These institutions played important roles at critical times, for example, during cholera, typhoid fever, and yellow fever epidemics.[38]

The port of Santos, in São Paulo, was evolving into a major trading hub, but poor sanitary conditions hampered work there. Frequent quarantines caused serious confusion, and foreign crews were leery of catching diseases. The government was alarmed when a large number of rats died in Santos in 1899. Two scientists called in from São Paulo's Bacteriological Institute, Lutz and the physician Vital Brazil, were quick to recognize the onset of a plague epidemic. Oswaldo Cruz, only twenty-seven years old and just back from Europe, was commissioned by the federal government to issue his professional opinion. Effective biological control measures were adopted to cope with the crisis, and prominent institutions for researching and making serums and vaccines were created in the epidemic's aftermath. The state began to rely heavily on microbiology in its public health policy.[39]

The government of São Paulo purchased the Butantan farm and equipped a laboratory to make antiplague serum there. First attached to the Bacteriological Institute, the laboratory became an autonomous institute and was renamed the Serum Therapeutics Institute in 1901, headed by Dr. Vital Brazil. It was the site of basic research in scientific medicine and also produced vaccines and serums. In a few short years, Vital Brazil would earn world renown for his production of anti-ophidic serus, or snake antivenom.[40]

In 1900, the Federal Serum Therapy Institute was established in Rio de Janeiro on the site of the Manguinhos farm, and Oswaldo Cruz was put in charge of operations. The institute started out as a laboratory for the production of plague serum but soon diversified to include research and instruction in a wide array of disciplines, while also building relations with the fields of public health, veterinary science, and botany. It became a center of excellence in the study and prevention of parasitic diseases and the home of research in serology, hematology, bacteriology, parasitology, entomology, and anatomic pathology. Manguinhos—as it was popularly known—was a "kindergarten of the sciences," in the words of Director Cruz.

From 1908—when the Federal Serum Therapy Institute was renamed the Oswaldo Cruz Institute—to 1922, the institute organized nine scientific expeditions to regions such as northern Minas Gerais and the Madeira, Amazonas, São Francisco, and Tocantins River valleys, as well as to a number of states in the Central-West and Northeast, blazing trails into Brazil's sertões. Reports were issued by several of the missions, but the one that garnered the most attention was written by two physicians, Belisário Penna and Arthur Neiva. Released to the mainstream press in 1912, their report depicted Brazil's interior and its inhabitants as backward, abandoned, and diseased. In 1917, Penna launched a "medical crusade for the Fatherland." One year later, the Pro-Sanitation League of Brazil was formed and served as the catalyst for various endeavors to "redeem" the people of the sertões; in the thinking of the sanitation movement, these people, reviled by many as inferior, could at long last be rescued from their frailty and infirmity thanks to public health campaigns.[41]

In a series of essays published in *O Estado de S. Paulo* in 1918, Monteiro Lobato declared that "the microscope had spoken," lifting a veil that had screened the country's biggest problem. Now the causes of the endemic diseases that raged across the land were visible, he explained: "Seventeen million creatures live to be used and enjoyed by the hookworm; three million pay stiff taxes in blood, life, and intelligence to a miserable bedbug; ten million shiver from the consumptive fever of malaria." In a country with a population of some 25 million, these figures only make sense if many of the sick harbored all three illnesses. Added into the bargain were Hansen's disease (leprosy), syphilis, leishmaniasis, and tuberculosis, which were also endemic.[42]

At institutes dedicated to research and teaching and to making serum and vaccine, scientific medicine was formulating new practices, while public policies were being designed for inhabitants in both urban and rural Brazil. Corroborating the notion of "illness[es] as phenomena affecting a population," the focus was now not solely on epidemics but also on endemic problems. If sudden death was the brutal fate of those in certain areas of the country during outbreaks of plague, smallpox, cholera, or yellow fever, endemic diseases meant that "death was now something permanent, something that slips into life, perpetually gnaws at it, diminishes it, and weakens it."[43]

The battle against hunger, neglect, and endemic disease took place at the crossroads of scientific medicine and biopower, connecting microbiology, medicine, territory, and nation. The population issue was at one and the same time scientific, political, biological, and a question of power. Medical authorities responsible for public health measures and urban reforms at times seemed to act as a "fourth power of the Republic."[44] This issue was crystal clear to the men of the era, like Roquette-Pinto, who believed that the central problem of modern nations was population—not just in quantitative terms but above all in qualitative terms, an aspect vital to worker efficiency. Unpopulated, the land lay useless, for only people, imbued with intelligence and might, could instill things with their due value.[45]

ENTOMOLOGY AND AGRICULTURE

If the Pasteurian revolution and the discovery of microorganisms had lent momentum to scientific medicine, the late nineteenth-century discovery of how insects figured as vectors for countless diseases truly ushered in the "golden age of entomology." In turn, the identification of animals that were hosts for worms and microbes underscored the importance of zoological research. The problem of disease demanded knowledge of entomology, fauna, and ecology, and physicians and biologists teamed up to study the morphology, physiology, life cycles, and geographic distribution of vector hosts.[46] Tropical medicine, ecology, zoology, entomology, and zoogeography pooled resources, targeting not just human populations but populations of bacteria, insects, and mammals.

The rise of biology in Brazil was also attributable to the strategic importance of this kind of knowledge in a country whose economic mainstay was the export of agricultural commodities. The creation of the Paulista Museum, in 1893, and the reactivation of the Pará Museum of Natural History and Ethnography, in 1894, were both initiatives undertaken by state governments whose local economies depended on the era's chief exports: coffee and rubber.

The Paulista Museum got its start through the donation of a sizable private collection by Joaquim Sertório to the São Paulo state government. Hermann Von Ihering, an internationally admired researcher and member of leading scientific societies, was invited to head the new museum. Based on its initial collection, the museum seemed destined to mimic traditional natural history museums, but what Von Ihering had in mind was a specialized institution, in step with the experimental character of biology in his day, offering a privileged place for scientific research in a country that had no university courses for the study of nature. Guided by this ideal, the Cajuru Biological Station—the first of its kind in South America—was founded in 1909 as a natural laboratory for the investigation of animal behavior and plant development.

The government of the state of Pará first proposed establishing a natural history museum in 1861, but did not do so until 1871, under the name of the Pará Museum of Natural History and Ethnography. Still, the institution was practically a nonentity until 1894, when the state invited Emílio Goeldi to direct it. During the rubber boom, the city of Belém underwent urbanization and intense demographic growth, and an affluent elite who wanted its town to become the "Paris of Sun" poured generous funds into the museum, which then experienced some of its most vibrant years.[47]

Goeldi worked hard to fashion a new kind of museum, envisioning institutes that would boast laboratories and equipment, led by great scientists. He hired staff, refurbished facilities, published scientific journals, and made systematic additions to the museum collections—all in pursuit of creating a paradigm of a new kind of museum and in repudiation of the outdated notion of museums as warehouses of musty old things. As a reputed zoologist, Goeldi wanted to build something akin to an institute for experimental biological research, linked to other international science centers through exchange, specialized publications, and interaction among researchers.

The museum created a number of biological stations and built a zoo that housed myriad Amazonian species; it also had a botanical garden, whose highlight was experimental research on rubber trees. Goeldi launched initiatives to breed animals in captivity, such as egrets, a pet concern of his and the motive for the protest he had lodged with Pará officials. The museum's scientific production was primarily biological, represented by its two principal researchers: Goeldi himself, a zoologist, and Jakob Hübber (1867–1914), a preeminent botanist and expert on rubber plants. When Goeldi left the museum in 1907, Hübber stepped in as director and stayed until 1914, which coincided with a period of increasing economic hardship in Brazil prompted by stiff competition from Asian latex.[48]

In the realm of biological research for agricultural purposes, applied entomology played a major role. It was taught as a course at the Higher School of Agricultural Science and Veterinary Medicine, founded in 1912, but it was not until the early 1920s—when the coffee berry borer was detected in the Campinas region of São Paulo and a grim epidemic ensued—that applied entomology moved onto the national stage. Previously, Brazilian coffee growers had been enjoying a period of remarkable prosperity precisely because the borer had delivered a crippling blow to competing growers in Southeast Asia. At the first signs of an outbreak in Brazil, the state government asked the entomologists Costa Lima, of the Higher School of Agricultural Science and Veterinary Medicine, and Arthur Neiva, then head of the National Museum, to join a commission of consultants. Costa Lima had just returned from a long mission to the Northeast, where he had led the fight against the pink bollworm (*Pectinophora gossypiella*).[49]

State officials also called in the agronomist Edmundo Navarro de Andrade, who years earlier had warned of the danger presented by the coffee berry borer.[50] The commission was charged with studying the biology of the insect, ascertaining what regions had been affected, and coming up with solutions. Informational campaigns were organized, flyers distributed, and educational films shown widely in rural towns and even on some plantations. Under Neiva's leadership, the commission drafted a proposal to establish a permanent institute, a request that was initially vetoed by São Paulo's lower house but then approved in 1927. Thus was born the Biological Institute of Agricultural and Animal Defense, in the city of São Paulo, with Neiva as its first director.[51]

Neiva had headed the National Museum from 1923 to 1926, and during this period he had worked to modernize the institution in scientific terms. Experimentalism was then holding sway in biology research, and revisiting museum practices was a top concern. The previous director (from 1876 to 1893), Ladislau Netto, had already started moving in that direction. He had advocated the creation of laboratories and dreamed of opening a marine aquarium, botanical garden, and zoo; he had reshuffled the museum's departments, shifting research emphases and introducing new fields, like applied zoology and botany. In 1880, he founded a laboratory for experimental physiology, where research was conducted on toxic substances, particularly snake venom.

In 1909, the museum was transferred from the Ministry of Justice and Internal Affairs to the Ministry of Agriculture, Industry, and Commerce, with the intent of making it an advisory agency on technical matters. The laboratories for analytical chemistry, plant chemistry, phytopathology, and agricultural entomology were added in 1911 as a resource for agriculture. In 1922, during the centennial of Brazil's independence, an official museum publication stated that the institution no longer filled the "outmoded role of collector, closed to scholars and indifferent to the nation's economic development" but instead sought to contribute to progress, just like the "most diligent research establishments."[52]

It was in this context that Neiva took charge of the museum, enthusiastically pursuing the goal of adapting the institution's practices to the standards of cutting-edge research centers. His actions were limited, however, since much work was required to beat the coffee plague, and he also had to devote significant energy to setting up the Biological Institute of Agricultural and Animal

Defense. His successor was Edgard Roquette-Pinto, who, as head of the National Museum from 1926 to 1935, instigated a new phase in its activities.[53]

SCIENTIFIC EXPEDITIONS AND NATIONAL TERRITORY

Roquette-Pinto had previously been directly involved in an endeavor that helped revitalize biological research at the National Museum: the Strategic Telegraph Commission of Mato Grosso to Amazonas. Inaugurated in 1907, the initiative was part of a larger project to carry Brazil's republican nation into the interior; its strategy was to unify the country territorially by hanging telegraph lines to ensure communications and make nationwide surveillance possible. Yet the work of the scientists on the commission was no less important; indeed, it has been said that "scientific exploration was one of the goals of the Strategic Telegraph Lines"—and a strategic one at that.[54] João Batista de Lacerda, director of the museum from 1895 to 1915, repeatedly called for reinstatement of the post of traveling naturalist—abolished with the Proclamation of the Republic—arguing that fieldwork was crucial to replenishing the museum's collections. Brazilian president Afonso Pena granted his request, and the museum joined the telegraph commission in 1908, surveying and studying the flora and fauna where the expeditions ventured. At a time of national efforts to painstakingly assess Brazil's available natural resources, its agricultural potential, and its animal resources, as well as to measure the dangers posed by poisonous animals and insect vectors of tropical diseases, these territorial explorations spawned numerous investigations, with biology at the forefront. Through its collaboration with several of the expeditions, the museum invigorated its collections with new material while its scientists broadened their knowledge. A number of its top scholars in biology took part in these journeys, including Alípio de Miranda Ribeiro, Frederico Carlos Hoehne, João Geraldo Kuhlmann, and Alberto Sampaio.[55] From 1908 to 1916, the institution received 17,886 new items: 8,827 botanical, 5,637 zoological, 42 geological, and 3,380 anthropological. These additions breathed new life into the collections and generated a multitude of studies and publications.

The scientists' experiences in far-flung corners of Brazil were pivotal in intertwining the construction of biological knowledge with the construction of interpretations of Brazil. Telegraph lines were the threads of the republican government's strategy to establish its presence in northwestern Brazil and thus allow for the incorporation of isolated lands and populations. This would not be done without exposing enormous contradictions within Brazilian society, however.

The government had appointed Cândido Rondon to lead the telegraph enterprise (which has become commonly known as the Rondon Commission). Rondon had been a republican activist in the army under the empire and was a fervent positivist. While working on the lines, he spent long stretches of time in the sertões and forests of Brazil, where he made contact with indigenous peoples living in complete isolation. He fought to protect them and defend their land against the local rubber barons and traders. In 1910, the Indian Protection Service began official operations with Rondon as its director. He and the other expeditioners lived tense moments in the interior of Brazil; their meetings with indigenous peoples and the rural poor revealed how hard it would be to enforce republican practices across Brazil's vast territory, where private interests wielded unbridled power.

In 1914, Roquette-Pinto accompanied one of Rondon's expeditions as a researcher with the National Museum, an experience that would later prove decisive in his work as director of the museum. He was struck by the violence and lawlessness that reigned in the regions they traveled. In Corumbá, in the state of Mato Grosso, he was told that people there had their own law for resolving conflicts: "article 44, paragraph 22"—an allusion to the caliber of a rifle and a handgun.[56] Reminiscent of the warning sounded in reports by the Manguinhos scientific missions, Roquette-Pinto's account of his trip describes a forsaken people, consigned to misery and illiteracy, their social relations grounded in selfishness and violence—the counterpart of a weak state and the consequent need for regular, ongoing attention.

In tandem with these endeavors, there was a growing awareness that a notable community of Brazilian scientists was taking shape. Scientists like Oswaldo Cruz, Carlos Chagas, and Vital Brazil won major international awards. As Mello Leitão was to write in 1937, Manguinhos had shifted world opinion about Brazil, and its scientists were now recognized for their skill in understanding and resolving scientific issues. Men of the stature of the physician and botanist Francisco Freire Alemão had previously been asked by foreign naturalists to ship them material for classification in Europe, but it was a new day now and scientific exchange took place on an equal footing. A slew of institutions solicited Neiva's opinion on Triatomine bugs, Costa Lima received flies from around the world for analysis and research, Lauro Travassos was invited to teach at the University of Hamburg, and Mello Leitão himself, an arachnid expert at the National Museum, networked with institutions in Barcelona, Buenos Aires, La Plata, and Montevideo. In short, Brazilian scholars of biology won prizes, received invitations, and were invited to add their perspective to international debates as peers and even as mentors.[57]

In sum, as Brazil struggled with its great problems in the early decades of the twentieth century, biological knowledge was developing hand in hand with the country's strategies for national salvation. Biology informed eugenic arguments during the debate over the alleged inferiority of the Brazilian people. It explained that devastating epidemics were caused by bacteria and microbes. It pinpointed insects as vectors of endemic diseases, and animals and people as the hosts of maladies that left millions of Brazilians invalids. It clarified the connections between invertebrates and crop destruction. It traced out possible paths to building a renewed nation. It gained strength from the establishment of research institutes and the revitalization of existing natural history museums. It saw a number of Brazilian researchers propelled into the international spotlight because of the quality of their intellectual production. It guided public health and sanitation policy that affected both human populations and the animal and plant kingdoms. And in a feedback loop, just as the Rondon Commission benefited from the acquisition of biological knowledge—one of its goals—biology drew its own rewards from participation in the telegraph enterprise. In short, biology was a wholly strategic and, above all, political form of knowledge with the potential to help a nation and its people come into being.

To return to the original matter at hand: The audacity of a bill to regulate hunting, which challenged entrenched economic interests and practices and assigned a decision-making role to the scientists at the National Museum, must be understood against the historical backdrop of the early decades of the twentieth century. As I have argued, biological knowledge was playing a vital role in a number of the biggest controversies surrounding national projects. The actions of Roquette-Pinto, Mello Leitão, and Alberto Sampaio reflected an atmosphere of optimism. The very fact that they presented their proposal tells us they were confident that the suggested measures had a real chance of being implemented; indeed, their very assignment to the task by the ministry was an acknowledgement of their status as authorities. This of course is not to say that it was all smooth sailing for the scientists, who at times faced opposition or felt snubbed or underappreciated. The proposal to enact controls on hunting was in fact a steep wager—almost a bluff—in the tense political game being played out under Vargas's Provisional Government, a game in which various groups faced off in a fierce struggle to define the new directions the nation would take.

IN SEARCH OF PROTECTION

The role that biological science assumed in nation building in Brazil was not without profound paradoxes. At no level did the government display regular, predictable support for either scientists or their institutions. During the early decades of the republic, the flow of public funds to museums and institutes was very uncertain, as was the real status of the researchers in the eyes of the public. Given institutional hardships, tensions between the federal and state governments over competing jurisdictions, and the sharp social, economic, and cultural differences between the states, support for scientific activities rode a roller coaster of steep ascents and dives. Added to this was the political and financial volatility of the period, which saw major outbreaks of inflation, swinging commodity prices (including the 1912 collapse of rubber production in the Brazilian Amazon), internal battles in the sertões that prompted forceful responses like military campaigns, and expenditures on European immigration to rescue coffee from its labor shortage. Brazil was a poor, indebted country, and public funds went where they were needed most, oftentimes eaten away by corruption. In this thorny context, scientists and institutions—especially those less directly involved with pressing public health issues—often lacked support and were undervalued by society and public officials.

In a 1914 article on the need to protect birds, Hermann Von Ihering bemoaned Brazilian officialdom's disdain for nature and science. He felt the Paulista Museum exemplified this well; although the institution offered fine information services and extensive, well-studied zoological collections, few sought its assistance. Among these were public health physicians, "sometimes in need of biological information, sometimes inquiring about the classification of animals harmful to health or of interest because of their parasites or parasite transmission." Agricultural scientists also contacted the museum fairly frequently. But apparently neither the Chamber of Deputies nor the Senate was aware of any of this activity, since, the director complained, both houses fired off harsh and baseless criticism and even discussed the possibility of shutting the place down. Because of the lack of staff and money, the priceless natural treasures of Brazil stored at the museum were not properly preserved. In Brazil, the dominant attitude toward science was utilitarian; it was valued solely when its practical results were urgently needed to respond to some disaster or blight. Government action was necessary if the country's natural wealth were to be defended steadily and systematically, and Von Ihering argued that "it is biology that should guide us with its teachings."[58] In 1915, in the climate of World War I, Von Ihering became a victim of the animosity directed toward Germany: following a series of controversies over his administration of the Paulista Museum, his public opinions, and his dreams of founding a specialized mollusk institution, he was removed as the museum's director. He chose to leave Brazil and continue his career elsewhere.

The decline in the rubber trade also spelled a rough period for scientific work at the Pará Museum of Natural History and Ethnography. In 1907 Goeldi went back to Europe. He was replaced by Hübber Snethlage, who passed away in 1914 and was succeeded by his wife, Emília Snethlage. As funds dried up, the place fell into a state of neglect that was much lamented by Brazilian biologists of the day. The zoo wasted away, its birds were sold to an institution in New York, and precious pieces were destroyed by insects. A sad incident occurred in 1921: the governor of Pará reprimanded Director Emília Snethlage for diverting food meant for the zoo animals to the impoverished staff.[59]

Research institutes received a great deal more attention than museums, however. Not that they suffered any less from the vicissitudes of public policy: pressing needs were attended to in times of crisis, but these responses were interspersed with long periods of government neglect. Arthur Neiva had met with crushing defeat the first time he tried to establish the Biological Institute of Agricultural and Animal Defense, in 1926; he decried his failure as a clear sign of government disregard for science.[60] Despite Vital Brazil's resounding success with his own research and other initiatives, the Butantan Institute battled mightily for funding. The institute distributed almost all of its products for free, even when demand rose. Vital Brazil's repeated requests for authorization to market the products fell on deaf ears and he left the institute. Although this authorization was eventually granted, the situation worsened after Vital Brazil's departure, and subsequent directors complained repeatedly about government disregard for the institution. In the early 1920s, Butantan lost researchers, its journal *Memórias do Instituto Butantan* came out only irregularly, services grew disorganized as many technical personnel suddenly left, and research was cast aside in favor of the simple production of serum, vaccine, and drugs. There was not enough money to buy corn and forage for the animals that were kept on hand to be bled for serum, nor was there any way to import indispensable laboratory materials.[61]

Manguinhos went through rough times as well. The 1920s saw a colossal shortfall of funds. Money earmarked for research was used to pay for basic overhead and infrastructure, and as a result, equipment became obsolete, facilities deteriorated, researchers' wages declined, and—mirroring the trend in other research institutions—staff began working two jobs. Some scientists, "taking an initiative that was unprecedented in the conduct of the austere and conservative members of that community," publicly denounced the disdain shown by government authorities and the decay of this celebrated research institution.[62]

In view of all this, even though biology was seen by intellectuals and policy makers as providing knowledge essential to nation building and strategic to political initiatives, scientists shared a general feeling that they were undervalued by public officials and society at large. The Brazilian people had been debilitated by endemic disease and decimated by recurring epidemics, the country's agricultural economy was threatened by devastating plagues, and its environment was being eroded by dubious private interests. Consequently, the idea that Brazil's citizens, agriculture, and natural heritage had to be protected was invariably accompanied by proposals based on the notion that a strong, central state—motivated by reason of state and guided by the collective interest—was the agent best geared to pave the way for realizing the nation and electing scientists as the guides of this process. And scientists wanted to be protected too.

NEW ERA, NEW HOPES

In the immediate wake of the so-called Revolution of 1930, new hope was ignited that the state would do something to constrain Brazil's oligarchic elites and the economic interests working in opposition to the best interests of the nation's future. In the case of the hunting bill written by the three members of the National Museum, the hope was that the Provisional Government would redress the scientific community's longstanding grievances about protecting Brazilian fauna.

The bill had been drafted at the request of the MESP, then led by Francisco Campos, leader of a major educational reform in the state of Minas Gerais in 1927. Established in November 1930—almost immediately after the revolution—this ministry figured strategically in the Provisional Government of Getúlio Vargas, whose political project was markedly centralizing in nature.[63]

The protection of nature was thus initially framed as an educational matter, and this perspective dovetailed with the ideas espoused by the drafters of the bill. Mello Leitão, in the interview mentioned earlier, posited the protection of natural resources as the true index of civilization and culture. Likewise in 1932, during a course given at the National Museum under the auspices of the University of Rio de Janeiro, Sampaio cited examples of countries whose education ministries had special departments that were entrusted with protecting nature and that advocated instruction in the topic at all levels of schooling. He highlighted the link between nature and health and argued that appreciating and safeguarding flora and fauna were the best ways to prevent the starvation and disease caused by substandard living conditions in rural Brazil.[64]

The ministry's request for the bill came during a period when Brazil had no constitution in effect and the country was ruled by so-called decree laws. The bill's authors fervently hoped that a stronger state, following Decree 19.398, signed on November 11, 1930, which placed both executive and legislative powers in the hands of the head of the Provisional Government, would be able to rein in powerful and entrenched interests and practices. The text of the bill argued that its stringency was justified for the sake of the greater good of the nation. Still, it was open for suggestions from any Brazilian citizen for three months following its publication on November 22, 1932.

It was a particularly tense time in Brazilian politics, involving redefinition of the country's institutional direction. With many projects up for discussion and multiple sectors of society mobilized to back them, "the government was operating in quicksand."[65] A sizable contingent, comprising young politicians and military officers who were members of the *tenentismo* movement, wanted the Provisional Government to be extended so that more radical nationalist, interventionist measures could be implemented in order to dismantle oligarchic powers and guarantee the ideals of the revolution. This same group also wanted to install technical councils at different decision-making levels to ensure that the central concerns of national reconstruction would be addressed in a nonpartisan, apolitical manner. Another opposing position, which included the São Paulo elites defeated by the Revolution of 1930, hoisted banners of a more liberal hue and endorsed federalist practices and the return to a constitutional regime. In May 1932, Vargas decreed that a commission be designated to draft a constitution, while he concurrently postponed elections for another year, to the immense dissatisfaction of many. The Constitutionalist Revolution broke out in São Paulo in July of that same year. The movement was vanquished three months later, but not before securing the government's promise to steer the country back to a constitutional regime. In May 1933, elections were held for the Constituent Assembly. Members were sworn in on November 15, 1933, and the definitive charter was enacted on July 16, 1934.[66]

It was during this period of peak political flux that Mello Leitão, Sampaio, and Roquette-Pinto were writing their bill. The tenentismo movement was still highly influential, championing a strong, central, authoritarian revolutionary government that, with the aid of nonpartisan agents organized in technical councils, would implement decisive, even daring measures for the nation. The bill composed by the three scientists undeniably had much in common with this vision, given that it proposed strict, centralized measures and rules, backed by a council of National Museum members who, in theory, were in tune with the real interests of the nation. The deadline for lodging amendments to the bill was February 1933, prior to the elections for the Constituent Assembly, and the idea of rule by decree still enjoyed broad support in the halls of Brazilian politics. In a context of clashing political projects and uncertain courses, the struggle to protect wild fauna was just one of various initiatives by scientists at the National Museum seeking the government's protection in their efforts to forcefully influence the building of the nation.

THREE STORIES: ROQUETTE-PINTO, SAMPAIO, AND MELLO LEITÃO

Writing the bill on hunting legislation was not the only joint project of these three scientists, and maybe not even the most important or the one with the biggest potential. If we compare their biographies through 1932—the year they composed the draft law—we will note that they worked together on a number of other common fronts and their stories were peppered with shared experiences. Rather than examining their backgrounds individually, let us take a look at how they developed their work and their identities as scientists within a web of mutually reinforcing activities.

All three had similar professional training and worked at the National Museum, where the staff was relatively small. In 1932, with Roquette-Pinto as director, the museum had eight professors who served as department heads—including Sampaio, head of botany, and Mello Leitão, head of zoology—plus two assistant professors and two traveling naturalists, in addition to technical personnel.[67]

All three had attended the Rio de Janeiro School of Medicine around the same time: Roquette-Pinto graduated in 1905 and Mello Leitão in 1908, while Sampaio had enrolled in 1903 but dropped out in late 1904 to apply for a position as a specimen technician at the National Museum. He later received his diploma as a homeopathic physician from the Hahnemannian Institute School of Medicine and Surgery, founded in 1912. From 1880 to 1889, their shared alma mater had undergone a sweeping reform that introduced new disciplines and replaced the previous emphasis on theoretical training with practical training. During their years of schooling as physicians, Brazil was feeling the influence of the new European tendency to prize laboratories and biological knowledge, a response to Pasteurian theory. Other changes came in 1891 and 1901, when courses were reorganized at Rio's School of Medicine and new attendance rules put in place. At the dawn of the new century, however, while all three were studying there, the school hit a rough patch; internal crises left the institution without a board of directors for a significant period of time, budgets were tight, and laboratories suffered neglect. Recalling his college days, Mello Leitão had harsh words to say about the classes given by Professor Joaquim Pizarro, who speechified about "life and death, Haeckel's monera," Lamarckism and neo-Lamarckism, pure Darwinism or Darwinism painted with the brush of Wallace, Haeckel, or Weismann, in a "barrage of verbal pretense." Pizarro also required students to read Darwin's *On the Origin of Species*, Haeckel's *The History of Creation*, and Albert Dastre's *Life and Death*.[68] Roquette-Pinto and Sampaio probably read the same works. For Mello Leitão, Professor Miguel Couto was the instructor who left the most lasting impression; Couto was also one of the members of the committee that judged Roquette-Pinto's medical thesis (a final requirement for completion of the course) while he was working as an intern in the professor's laboratory.[69]

Mello Leitão had been born in Paraíba, but his family moved to Rio de Janeiro when he was a child. He began his career as a zoologist in 1913 at the Higher School of Agricultural Science and Veterinary Medicine in Rio, which was forced to shut down in 1915 for lack of funding. At the end of the same year, he moved to Belo Horizonte, where he taught a class in pediatric medicine and child hygiene at the School of Medicine. When the Higher School reopened its doors in 1916, he returned to Rio and served as its head until 1919.[70] In 1916, he also spent a semester at the Rio de Janeiro Normal School, where he worked with Roquette-Pinto, who taught natural history there. The principal was Afrânio Peixoto, an intellectual devoted to building a modern, civilizing school attuned to issues of hygiene and its biological basis. Peixoto introduced the practice of rigorous qualifying examinations for candidate professors and the requirement that their résumés show training in a specialized discipline. Many of these practices, like the qualifying exams, were discontinued shortly thereafter but resumed again under the educational reform pushed through by Fernando Azevedo in 1927.[71] During this period, multiple sectors of Brazilian society were absorbed in debating the question of education, then identified as one of the most serious issues on the national agenda. Condemning the dreary way these subjects were presented in the classroom, Mello Leitão called for changes in the teaching of natural history and general biology. He advocated lively, dynamic pedagogical methods, with less theory, more practice, abundant classroom materials, and a prime focus on the study of Brazilian flora and fauna. He was an assiduous participant in the debates held by the Brazilian Education Association (ABE), formed in 1924.[72]

From 1923 to 1930, Mello Leitão taught natural history at the Rio de Janeiro Normal School and the Niterói Normal School. Noticing the large vacant lot next to the latter, he contacted Alberto Sampaio, by then at the National Museum, and together they planted a teaching garden and gave practical classes outdoors.[73] The educational material he produced during this period also reflected a penchant for experimental teaching: to encourage his student readers to investigate their own environments, Mello Leitão interspersed 456 prints of common Brazilian plants in his *Compêndio de botânica* (Compendium of botany), published in 1924. The textbook won critical acclaim, particularly for its clarity and method.[74]

In later years, Roquette-Pinto remembered as a student witnessing the first concrete steps toward creating a course centered on experimental medicine, when Professor Miguel Couto ("a full-fledged pioneer of this renewed art") inaugurated a "modest laboratory for Roentgen diagnosis at his infirmary's service." It was there that Roquette-Pinto and some classmates saw their first vial, in 1903. "It was also there," he related, "that for the first time, in a dark corner, a radioactive substance sparkled before our eyes, well before regular application of Curietherapy had begun."[75]

Roquette-Pinto was a founding member of the ABE and shared Mello Leitão's dislike for traditional teaching and his advocacy of urgent change. He started his career as a professor at the National Museum in 1906 and was a founding member of the Brazilian Academy of Sciences (ABC), established in

1916 as the Brazilian Association of Sciences. Sampaio and Mello Leitão joined the ABC in 1917 and were distinguished figures in the association in subsequent years.[76] Roquette-Pinto took part when the ABC organized Brazil's first radio station—Rádio Sociedade do Rio de Janeiro—and served as its secretary general. The greatest aspiration of the staff and supporters of Rádio Sociedade was to transform radio into a classroom accessible to Brazilians in the most remote corners of the country. Roquette-Pinto and the entomologist Costa Lima hosted daily programs.[77] Sampaio was also active at Rádio Sociedade from its earliest days and presented a number of broadcasts on the flora of Brazil. He joined the National Museum as a professor of botany in 1912 and was likewise an active member of the ABE, attending meetings and debates. Along with other scientists, Sampaio and Roquette-Pinto welcomed Albert Einstein when he visited both Rádio Sociedade and the National Museum during his 1925 trip to Brazil.[78]

Roquette-Pinto and Sampaio had something else in common: their travels with Cândido Rondon. In 1910, Roquette-Pinto spent several months studying collected materials with Rondon. The scientist said the voice of the "master" was like the "inviting voice of the sertão," conveying the "rustling of distant forests" and infusing his thoughts with "the poetry of those distant lands." But since Roquette-Pinto was in Europe in 1911, it was not until the next year that he was able to satisfy his desire to accompany Rondon on an expedition. The journey left a deep impression, in part thanks to his meeting the Nambiquaras, indigenous people "secluded in the heart of Brazil." In 1917, he published an account of the expedition, entitled *Rondônia*, in honor of Rondon. His purpose, he declared, was to "document and disseminate," "record and serve," because while he had been fortunate enough to encounter an experienced guide, many would not be, and through his story they might get to know a Brazil previously unknown even to Brazilians themselves.[79]

Sampaio spent 1909 working hard on a major botany collection that had been gathered by other naturalists in the region of Cáceres, Mato Grosso. His chance to accompany Rondon on one of his expeditions did not come until 1928, when they traversed the Tumucumaque mountain range between Brazil, French Guiana, and what is now Suriname. He researched the region's flora and published his results in a number of scientific journals; like Roquette-Pinto, he too published an account of his trip with Rondon.[80]

Roquette-Pinto cited the Uruguayan José Enrique Rodó in voicing his own thoughts on how travel shapes people: when men are held prisoner to their homeland, they suffer from "expanded solitude," condemned to live and die cloistered inside their own customs, thwarted from broadening their empathy—after all, "living means renewing oneself."[81] During his youth, Roquette-Pinto was so fascinated with travel that he had thought about enlisting in the navy. After he graduated from medical school, he took his first trip abroad, attending the Universal Races Congress in London in 1911. He stayed on in Europe, furthering his knowledge of anthropology and biology, attending conferences, and taking classes with physiologists, zoologists, and anthropologists—such as Charles Richet, who would win the Nobel Prize in Medicine in 1913; Alexandre Brumpt, a parasitologist and entomologist specializing in tropical insects; and Felix Von Luschan, who developed a chromatic scale for classifying human skin color. In 1920, he spent some time in Asunción, Paraguay, as a guest lecturer in physiology at the School of Medical Sciences. In 1924, he traveled to Sweden to represent Brazil at the International Congress of Americanists and there met the anthropologist Franz Boas. He also took the opportunity to revisit Europe's scientific centers. Soon after, he traveled to the United States.[82]

Sampaio's and Mello Leitão's lives and careers were likewise filled with travel. In 1913, soon after Sampaio became a professor of botany at the National Museum, he was assigned to study in Europe, and while there he visited botanical institutions and attended a number of courses. In 1926, he attended the World Forestry Congress in Rome, and in 1931 he represented Brazil at the Fifteenth International Geographical Congress in Paris. Over the course of these European travels, he involved himself in activities at a dozen botanical gardens and eleven museums, something that would leave an imprint on his writings: his books and papers always referenced what he had learned and the practices he had observed at these institutions. His exploration of the intersections of botany and geography prompted him to specialize in the phytogeography of Brazil and led to the 1934 publication of his book *Phytogeographia do Brasil*.[83]

Mello Leitão first visited Europe in 1926, in his capacity as director of the Children's Polyclinic in Rio de Janeiro, before joining the National Museum. In Paris he did an internship at the Hôpital des Enfants-Malades, but his afternoons were devoted to studying zoology at the Musée d'Histoire Naturelle, where he met the zoologists Charles Gravier and Louis Fage and the distinguished arachnologist Lucien Berland. He also visited a few elementary schools around Europe to take notes on the study of natural history and biology, both as president of the ABE and as a professor at the normal schools in Rio de Janeiro and Niterói. Right after he was hired at the National Museum, the minister of education and public health appointed him to represent Brazil on a scientific exchange trip to Uruguay in 1931.[84]

Through their travels, all three men forged ties that kept them in contact with international societies for the defense of nature and with numerous scientific associations around the world, including museums and universities. As members of these global networks, they maintained an extensive correspondence, as attested to by the papers stored in the National Museum archives. On the basis of these foreign contacts and experiences, all three could present themselves as scientific authorities of international renown, further empowering them to offer national policy suggestions. They staked claim to a distinctive understanding of Brazil—knowledge of its flora, fauna, and people—while they also kept abreast of trends in other countries and set standards in the realm of culture and civilization.

All three grounded their authority first and foremost in their knowledge of biology. Although Roquette-Pinto is remembered as an anthropologist (as well as a radio broadcaster, educator, filmmaker, and writer), it was his view that anthropology is a branch of biology. When it came time to choose a career, a friend of his father's had advised him to go into medicine, a suggestion he accepted because of how useful "a good course in the biological sciences" would be. His decision to take up anthropology in college was inspired by his classes in anatomy, which awoke him to the "interesting science of the human races." In 1926, he taught an anthropology course at the ABC; its syllabus leaves no doubt about his distinctly biological outlook on anthropology, with its classes on morphology, biometrics, genetics, cytology, comparative anatomy, comparative physiology, and paleontology.[85]

ROQUETTEA SINGULARIS AND *TYTIUS SAMPAIOCRULSI*

Despite the tremendous challenges that the field faced in Brazil, Roquette-Pinto felt biology had experienced an "admirable growth spurt" from the late nineteenth century to the early decades of the twentieth. An unwavering devotee of Alberto Torres, he underscored that thinker's dedication to the study of biology, arguing that it was this knowledge that had singled Torres out among so many other intellectuals of his generation who were adrift in the "glitter of discourse." Torres was an intellectual, politician, and jurist whose works exerted a heavy influence on Brazilian nationalist thought and who contended that a strong, central state was needed as an agent to act above the fray of conflict and as the organizer of the nation.[86]

According to Roquette-Pinto, Torres's library was "the most profound and finest that was to be had in the realm of biology," and when he died, he had left behind the latest publications ordered from Europe, their pages never leafed through. The head of the National Museum asserted that if Torres's work on social issues was seminal, it was only because he had "a sound foundation in biology."[87] Following in his "master's" footsteps in both ideas and practice, Roquette-Pinto habitually relied on biological concepts. Even when he ardently disputed theses that blacks, indigenes, people of mixed descent, the Japanese, and other immigrants were inferior, he did so "without discarding the presuppositions of physical anthropology and the foundational explanations of biology," in the words of Lima and Sá.[88] He denounced the flimsiness and scientific inconsistency of criticisms of miscegenation and rejected racism, without ever abandoning the concept of race or certain notions from eugenics.[89]

In 1932, Roquette-Pinto was one of those who signed the Manifesto dos Pioneiros da Educação Nova (Manifesto of the Pioneers of New Education), which contained repeated references to biology. Published by major news outlets on March 10, the document was "released in the very midst of disputes over the conduct of policies by the recently created Ministry of Education and Public Health in Brazil." A watershed moment for educational renewal in Brazil, the document called for the drafting of a national plan for free, compulsory, secular public schools that would offer the same education to boys and girls alike, from the ages of seven to fifteen. Despite its collective allusion to "pioneers" in its title, the document spoke for a rather eclectic mix of individuals, all of whom were in accord about the need for educational transformation. The manifesto itself was not wholly without precedent, since the early decades of the republic had seen a number of innovative initiatives along these lines, but it was "a political piece within the educational debate" of the day.[90] The manifesto cites Alberto Torres three times, while the term "biological" appears eleven. According to the text, education has a "biological nature" in that it is aimed at a population; it subordinates itself to the "biological functions" of the collectivity; it responds to the "biological right" to a well-rounded education; it guarantees the "biological purposes" of education (free, compulsory, secular); and it favors the differentiation of skills through "biological action." Biology is referenced to the collectivity here, not the individual, and the document evinces the ideas of biopolitics when it frames population "as a political problem, as a problem that is at once scientific and political, as a biological problem and as power's problem."[91]

At the National Museum—headed by Roquette-Pinto as of 1926—a number of overlapping initiatives had the tenor of political strategies. Because the institution was deeply involved with initiatives for educational renewal and was politically engaged with the regime in power, through its more active members it had its hands in myriad endeavors, including radio, Rondon's travels, the ABC, establishment of the ABE, educational film, the Pioneers Manifesto, international science conferences, editorial projects (like the *Revista Nacional de Educação* and the Brasiliana Collection), the formation of nongovernmental organizations like the Society of the Friends of Alberto Torres and the Society of the Friends of Trees, the 1934 organization of a national congress for the protection of nature in Brazil, and the proposal of related laws.[92] Within the country's new political context, dynamic and energetic scientists on the National Museum staff—like Roquette-Pinto, Sampaio, and Mello Leitão—carved out niches for themselves in countless strategic spaces, relying on a gamut of resources to do so and fighting to secure a unique place for the practice of science and above all of biology.

Over the course of the 1930s, Mello Leitão earned recognition as the most prolific arachnologist in Latin America; he described many new species and published 198 papers in specialized science journals from 1915 to 1951.[93] In 1931, the year he came on board at the National Museum and began his fruitful work alongside Sampaio and Roquette-Pinto, he paid tribute to his colleagues in the fight for biology by naming a spider *Roquettea singularis* and a scorpion of the family Buthidae *Tytius sampaiocrulsi*. Those were days of big investments and high hopes. In a way, ascribing such names to species in Brazilian nature was emblematic of the willingness and indeed yearning of these scientists to play a decisive role in defining the nation's future.

REJECTING CONFLICT

In September 1933, the *Revista Nacional de Educação*, with Roquette-Pinto as editor, published a commemorative first anniversary issue and distributed 15,000 copies free throughout Brazil. Featuring prominently toward the front of the issue was a speech by President Getúlio Vargas delivered at a ceremony in Salvador, Bahia. The journal entitled the text "Education," no doubt the theme to which it wanted its readers to pay special heed.

Vargas hailed the state of Bahia as the birthplace of nationhood—the place where the cross had first been planted by the Portuguese discoverers of Brazil on April 26, 1500, uniting the New World to Christian civilization and establishing the foundational pillars of both the nation that already was and the great fatherland that was yet to be built. According to Vargas, Bahia brought to mind the earliest efforts to tame this savage land and transform it into a source of wealth but, alas, the state also recalled the suffering of slave labor, an indefensible error committed by the empire with catastrophic consequences. Broad areas of land had been devastated by the "greed of the slave masters," who exploited the labor of "submissive," uneducated slaves. According to Vargas, at the outset of the republic destitute people roamed Brazil's rural lands; they were "sometimes almost nomads, living from hand to mouth," practically "anachronistic vassals of a feudal estate." A rural exodus had produced a miserable proletariat, the victim of privileged urbanites, who in turn had been born of an "aristocracy of the lettered elite" and benefited from the comfort of "government jobs [and] a life of pomp and lavish idleness." Vargas wanted to guide Brazil "back to the good path" by promoting a return to the countryside and by improving both public health conditions in the interior and the education of rural populations through elementary and technical schooling. It would therefore be up to the state to "lend the *sertanejo* [inhabitant of the sertões] . . . an awareness . . . of his rights and duties, fortify his soul and convince him that human solidarity does exist, and strengthen his body through hygiene and labor."[94]

This was not the first time the *Revista Nacional de Educação* had published texts signed by top officials of the Provisional Government. In its debut edition, Francisco Campos had introduced the magazine as "the first federal contribution to the work of educating the Brazilian people."[95] The periodical clearly supported the political interpretation that 1930 had represented a rupture with the "Old Republic"—"liberal, oligarchic, weak, inept, Europeanizing, and at a political and cultural remove from the 'Brazilian people.'"[96] Campos also sang the praises of Roquette-Pinto, whose skill and tenacity embodied "a sure guarantee that the magazine is a work destined to endure and succeed."[97] The magazine positioned itself as a contact point between the jurist Campos—responsible for the educational reform in Minas Gerais in the late 1920s—and the scientists at the museum. Campos and the museum staff also shared an affinity with the Escola Nova movement; both were active participants in the ABE and had a common political vision that tended toward bureaucratic centralization and the standardization of education procedures nationwide.[98]

The same issue featured the inaugural speech by the new minister of education and public health, Washington Ferreira Pires. Campos himself had stepped down as minister just before the first issue hit the stands. Perhaps the publication of his successor's speech in the opening pages of the premier edition was part of a strategy to ensure his support for the fledgling periodical. Be that as it may, in his speech Pires addressed subjects dear to the magazine. He declared his loyalty to the guiding principles of the Vargas administration and his staunch pledge to "heal the body and refine the spirit—to eugenize and instruct," all in an effort to prepare the Brazilians of tomorrow. Trained as a doctor, he underscored the various battles being waged against the maladies then afflicting the Brazilian population, like yellow fever, smallpox, the bubonic plague, Hansen's disease, malaria, venereal diseases, and Chagas disease. He described the government "as a complex living being with a well-organized structure," damaged every time one of its prerogatives was diminished or any of its elements torn away.[99]

The speeches reproduced in the pages of the magazine hold keys to understanding the rapport between scientists from the National Museum and the public power structure put in place following the Revolution of 1930. Campos pressed home the idea of educational initiatives firmly organized by the state. Pires used the image of a living organism, whose working parts had to be kept in balance. Vargas evoked images of forsaken rural poor and destitute laborers, dupes of the selfish elites, whose dignity had to be restored through education and hygiene under the guiding baton of a protective state so they would learn to trust in human solidarity again. The scientists at the National Museum endorsed the same ideas, which were wholly in keeping with their own views of the importance of biology. The fact that these men of science and the new administration had certain perspectives in common helps account not only for this general rapport but also, specifically, for the privileges accorded the former, like official financial backing for the magazine and the appointment of these scientists to write bills.

The notion of an organic whole was a consensual cultural allegory of the nation, state, and land, where the social classes were represented as interconnected organs working together homogeneously and harmoniously. As Alcir Lenharo has pointed out, the metaphor of "the body . . . rich in political implications" has surfaced time and again throughout history, but at this time it took on a special meaning and was embodied in the form of corporatism. Alberto Torres became a lodestar for intellectuals, politicians, junior military officers from the tenentismo movement, and scientists. In 1914, when he published one of his main works—*A organização nacional* (loosely translated, "Building a national organism")—he imprinted a particular political meaning onto the concept of organism.[100]

When Vargas, in his cited speech, laid emphasis on the danger of abandoning hinterlanders and laborers to the mercy of the rural elites and the bourgeoisie, he was reiterating a longtime concern of many. We should not underestimate the impact of rural millenarian movements like the Canudos War in Bahia or southern Brazil's Contestado Rebellion. In 1912, during a lecture at the Colmeia—a nationalist propaganda society in Rio de Janeiro—Roquette-Pinto warned that it was necessary to overcome the "extreme suggestibility" of the Brazilian people, "evidenced by the religious episodes that have cloaked forms of collective madness in Bahia and Santa Catarina." No less worrisome were the ascension of anarchism, the 1917 labor strikes, the 1922 founding of the Brazilian Communist Party (PCB), and the news of revolutionary movements worldwide. A few weeks after the birth of the PCB, Mello Leitão spoke out about the plight of the Russian people, whose ignorance and naïveté had left them susceptible to Bolshevist appeals and "wallowing in servitude, suffering, chaos." The Communist revolution had been a "devastating, catastrophic" event that fed "selfish and self-serving designs, veritable nihilistic, terrorist machinations." It was not by happenstance, according to Mello Leitão, that Russian scientists had fled after the revolution "like birds chased by a storm" or had died of hunger, been murdered, or sent to Siberia as "enemies of the proletariat."[101] For Roquette-Pinto and Mello Leitão alike, the idea that ignorance led people to succumb to both madness and revolt reinforced their vision of how education should figure large in nation building.

In 1930, the Provisional Government began enforcing measures to disband protest movements while simultaneously trying to tempt men and women from the countryside and from the factories with idealized images of a comfortable middle-class life and a healthy, hearty body ready for work. In this regard, the establishment of the Ministry of Labor and the MESP right after the Revolution of 1930 should be viewed as a decisive political strategy. The Ministry of Labor took up the mission of reconciling interests and preached appeasement of conflict through state action for the sake of the nation. The MESP supported education and hygiene as definitive solutions, sufficient unto themselves for redeeming the Brazilian people. Political conflict was thus exorcized as the perverse fruit of a neglectful liberal state that incited class resentments, which would be surmounted in a higher stage of collective life. The objective was to encourage "the replacement of the negative concept of class struggle with the positive concept of class collaboration," effectively depoliticizing society and, especially, workers. Through a series of gradually enacted labor benefits, labor unions were yoked to the state and bureaucratized under the management of technocrats.[102]

In the years following the Revolution of 1930, the scientists at the museum were swept up in the possibilities of social transformation, particularly because they felt they had the chance to lodge their grievances directly before public authorities, in the context of a new regime that was arrogating to itself the task of rebuilding the nation. It was all about reaching a stage in the country's evolution, much like an organism reaching maturity through full and harmonious functioning.[103] Change was urgently needed and could not be delayed; the list of tasks was as endless as the Brazilian territory, as diverse as its people, as profound as its tragedies, and as rich as its potential. The scientists wanted to do away with the primacy of personal interests, to which the true interests of the nation were being sacrificed—epitomized in the slogan launched by Alberto Sampaio as a cry for cooperation: "Nothing of selfishness; everything for Brazil!"[104]

Within this climate of sharp repudiation of the notion that conflict is an inherent component of social life, and in tandem with the defense of a corporatist, organic society, these scientists developed their positions on evolutionary theories as an integral part of their commitment to social and historical endeavors. For this reason, conceptions about life and evolution captured great attention.

THE WEB OF LIFE

The early decades of the twentieth century witnessed fierce rejection of Darwin in Brazilian intellectual circles. His theories had initially received a relatively warm reception in the closing decades of the nineteenth century. Intellectuals of the Escola de Recife movement debated natural selection and competition between species in the 1870s, as did attendees at the conferences sponsored by the National Museum during the days of the empire.[105] But in these milieus, Darwinism was mixed together with the theories of Haeckel, Spencer, and even Lamarck, creating a climate of marked eclecticism where theories with often conflicting scientific underpinnings were tossed together and various understandings of evolution stirred together in one big pot.[106] But as time went on, a growing rejection of Darwin accompanied the rise and eventual hegemony of an organic view of society.

The way Mello Leitão saw it, the initial success of *On the Origin of Species* could not be attributed to its scientific merits. Even though he ranked Darwin as one of the foremost scholars of the study of living organisms—alongside Aristotle, Cuvier, and Mendel—he was convinced that Darwin was thoroughly outdated. The spread of Darwinist thought was, in Mello Leitão's mind, due to three factors. First, "he appeared as the great ally of liberalism because he elevated the doctrine of free competition to the level of a natural law," which could explain why the elites acted out of self-interest, oblivious to their crushing the weak and laying waste to the nation's natural heritage. Second, Darwin's thinking served to justify the imperialist actions of aggressive peoples, allowing them to elect "survival of the fittest as a basic principle of their politics and a pretext for domination." In a context of burgeoning anarchist and communist labor movements, there was undoubtedly a third factor implicit in Mello Leitão's argument, and that was the radical materialism of Darwin's theory as well as its harmony with the notion of class struggle.[107]

Mello Leitão counterposed Darwin with the rediscovery of Mendel's theory, systematized in 1865 and revived around 1900 by the biologists Hugo de Vries, Carl Correns, and Erich von Tschermak-Seysenegg. Feeling that Mendel stood at odds with Darwin, Mello Leitão used genetics to ground his rejection of natural selection as a factor in evolution.[108] He did not dispute evolution itself or the ongoing transformation of living organisms and indeed believed in "the steady, incessant transformation of nature." But in his mind, a different mechanism accounted for the origin of species: genetic mutations.[109] Rebuking the idea of life as an arena for unending struggle, Mello Leitão—like so many others of his day—envisioned nature as a holistic ecosystem, a "web of life" in which groups and species were members of large community units, harmoniously sharing certain habitats. His overriding image of nature was one of a large organism in balance, comprising networks of interdependent links, grounded in mutual aid, altruism, and sociability. From this standpoint, struggle and competition were anomalies; in a vision of nature where species partook "of the sumptuous banquet of life," struggle was a "false, thankless word, which incites appalling misunderstandings." With evolution thus cleansed of its hard, cruel feature of an endless struggle for survival, it could be seen as a linear movement toward the perfection of interconnected beings within a harmonious whole.[110]

Roquette-Pinto likewise condemned the notions of the struggle for existence and the selection of the strongest and the fittest through the elimination of others. According to him, the scientists who followed Darwin were forgetting "the brilliant constructs of Lamarck," who had clearly demonstrated "the irrefutable influence of the environment as it propels the adaptation of organisms." He celebrated the return of Lamarckism to Brazilian intellectual circles and reaffirmed the hypothesis that species transform without any struggle, thanks to adaptation; this concept of evolution was quite distinct "from the fierce antagonism of early times" and from the conflict favored by pure Darwinism, while it was also the reverse of the determinist excesses committed by the eugenicists of his day. Citing the biologist H. S. Jennings, Roquette-Pinto waged war against the notion of an ineluctable genetic heritage and labeled the distinction between acquired and hereditary traits as artificial. He held that heritage was no more than a potential or a predisposition that was crucially dependent on environment: "Nothing is more *changeable* than so-called *hereditary* traits" (emphasis in original). In the case of human changes, the key process was education; knowledge and inventions had to be disseminated. Roquette-Pinto was an ardent defender of a vision of "humanity [as] extraordinarily changeable and thus refinable, so long as living conditions are altered." The advance of physiology, his thinking went, would one day show that apparently hereditary morphological traits were the product of biochemical factors. He also mentioned Mendel, whose work—contrary to the fatalistic determinism of a number of his contemporaries—showed how new species could emerge "without struggle, through mutation," and he cited lab experiments in which fruit flies modified by X-rays had passed on their acquired traits. From then on, he felt, the study of nature would reveal the reconciliation of all creation and relationships of solidarity among living organisms. In this context, studies of life would have "a very lofty purpose," for they explore an "integral history of nature," in turn making possible the true study of ecology, which is nothing more than the science of cooperation among all living organisms.[111]

There was, however, one exception to this interpretation of nature as a perfect web: humans. Left to their own devices, they would be guided by selfishness and competition, and so a regulatory element was needed to restore harmony and align humans with the rest of creation. The state appeared as this intermediary, posited not as the product of a social contract but as the metaphysical manifestation of an idealized—and above all organic—nation, whose various social groups would merge their interests for the sake of a higher truth. As we can see, there was a great deal of affinity between biological concepts, authoritarian political presuppositions, and nation-building projects.

For scientists like Mello Leitão, Roquette-Pinto, and Sampaio, forging ties with the government reflected their hope of receiving financial support for initiatives to popularize science, which were intended to promote an environment favorable to transforming ignorant people into civilized citizens, in the finest style of evolution from a neo-Lamarckian perspective. They were also excited about the possibility of influencing public policy as consultants in the writing of bills or as members of decision-making boards. And their tenancy at the National Museum was the springboard for devising these strategies.

For the Vargas Provisional Government, relations with these scientists had the potential of shoring up its legitimacy. This kinship would also reinforce organicist, corporatist projects meant to supplant social conflict and seed peaceful social transformation, all steered at realizing the nation and ensuring a progressive, teleological evolution. This partnership was neither guaranteed nor absolute, as demonstrated by the process of pushing through the hunting regulations, which dragged out from 1932 to 1934. It was a time ripe with new possibilities, where a range of actors threw themselves into the game of politics, intent on having a voice in the nation's new directions.

THE 1934 GAME AND FISH CODE

On January 2, 1934, Brazil's *Federal Register* published the Game and Fish Code in the form of Decree 23.672. The final version of the law attests to the involvement of the National Museum scientists in its drafting. First of all, the two points underscored by Mello Leitão in his interview found their place in it: the fostering of scientific knowledge and the valorization of Brazil's science institutions. The collection and transportation of the eggs and larvae of aquatic species were banned, except when duly licensed by the National Game and Fish Council, provided for under the decree and subordinated to the Ministry of Agriculture. There was also an effort to keep collections from leaving Brazil and to control what could be shipped abroad. Foreigners had to file research requests via their governments and institutions, provide detailed information on the nature and duration of their studies, and report their precise fields of research. Upon completion of their investigations, they were to submit a report with their conclusions and list all items collected, along with points of origin. They were prohibited from sending specimens of eggs, larvae, or animals abroad without official consent. Authorizations to do research or animal control work were free and automatic for Brazilian scientists, and the post of researcher at national institutes now carried greater weight. Enforcement would be exercised by the National Game and Fish Council, headquartered in Rio de Janeiro, whose eleven members would be appointed by the minister of agriculture. One of the eleven positions would always be held by someone from the National Museum.[112]

That same year, the government put other measures in place as well. The Forest Code was created by decree a few days later, on January 23, 1934. Although Sampaio had not been on the committee that drafted the bill, he had made a number of suggestions to the original text, not only as a botanist with the National Museum but also as a member of the Society of the Friends of Trees. The code defined the country's forests as of common interest to all Brazilians. It also provided for areas to be set aside for permanent protection if they displayed special features that characterized them as true public assets. This particular article laid the legal foundations for the creation of Brazil's first national and state parks. The code also instituted the Federal Forest Council, on which the National Museum would likewise be represented.[113] Further legislation included the Law on Scientific Expeditions, which covered missions of any nature within Brazilian territory, the Water Code, and a law that stipulated a number of animal-protection measures and covered domesticated animals as well. Through the Law on Scientific Expeditions, the Ministry of Agriculture enacted controls over botanical, zoological, mineral, paleontological, and historic specimens collected inside the national territory. It also regulated the Brazilian Council on the Oversight of Artistic and Scientific Expeditions, made up of seven members, two of whom were required to be professors from the National Museum.[114]

Yet if we compare the bill of November 1932 with the final decree handed down in January 1934, we will note one undeniable element of defeat. As we know, the bill had been written at the request of Francisco Campos, minister of education and public health until September 16, 1932. By the time the bill was published in Brazil's *Federal Register*, on November 22, 1932, Washington Ferreira Pires had assumed the post. But the final decree placed the matter under the auspices of the Ministry of Agriculture, then headed by Juarez Távora. So from the time the three scientists were appointed to draw up the bill until Vargas signed the final decree, the hunting question was shifted from the sphere of public policy on education to the more economic, commercial, and logistical realm of agriculture.[115] It would be wrong to say that the Ministry of Agriculture was opposed to the bill submitted by the scientists or that the final decree was unconcerned with the conservation of nature. However, in its final form, the law failed to adopt a number of the ideas contained in the initial bill and also granted the National Museum much less maneuvering room than it wanted. It should not be said that the scientists did not think the new code represented progress. Nevertheless, they must have been disappointed by the gap between their initial proposal and the actual code and, above all, by the meager power ultimately delegated to them.

Government policy thus did not fully embrace the projects advanced by the museum's scientists. A few years later, Mello Leitão expressed his chagrin over the situation at the museum. In an English-language copy of a book by Louis Agassiz, who had spent the year 1865–66 in Brazil, he ran across a comment that Emílio Goeldi had written back in the 1880s. In the highlighted passage, Agassiz described the museum (then under the empire) as antiquated: nothing was being added to its collections or done to improve them; its taxidermied animals were shabby; there were few specimens. In the margin, Goeldi had penciled in the words "still the same today" and signed his name. When Mello Leitão read those pages nearly half a century later, he concluded that, despite Roquette-Pinto's mighty efforts, Goeldi's words would likely still resonate with an impartial visitor: "still the same today." He held Goeldi blameless, because he had spared no effort in his attempts to modernize and energize the museum—the institution's print shop had released more publications, the film library had grown markedly, a projection room had been added, and a number of educational initiatives had been promoted. But the nation's representatives had failed to appreciate how useful the museum could be, and its work went unrecognized. Mello Leitão called special attention to the obstacles erected by the "second Minister of Education," who "deprived the museum of any means for continuing its work"—an allusion to the loss of the decisive political support it had received under Francisco Campos.[116]

When the code was released, it was not accompanied by any measures that might have ensured its enforcement. Brazilian hunting and fishing practices changed little, and the scientists continued to bear witness to the tragic fate of the country's wild fauna. In 1937, Mello Leitão took stock of the relations between Brazilian society and its fauna. Fashion trends still dictated the persecution of many species. Feathers were being used less, but chinchillas and otters paid with their lives to feed the market for fur coats and stoles. On the outskirts of Rio de Janeiro, butterflies were being decimated to adorn knickknacks, while birds were stalked for the sheer fun of it. "Deer, tapirs, wolves, otters, and beautiful butterflies" were vanishing from Brazil. Although there was now a Game and Fish Code as well as a Forest Code, both were ignored by woodsmen and hunters, deemed pointless as far as traders were concerned, and the butt of jokes for politicians, who laughed at the laws as if they were "books of sordid humor." It is worth noting that the museum chose Mello Leitão as the scientist to represent it on the National Game and Fish Council, and he would serve as its chair from 1935 to 1942. But the undeniable reality of the worthlessness of the new code in controlling illegal hunting was a source of consternation for Mello Leitão.[117]

In the early 1930s, however, the future was as yet undefined, possibilities were vast, and the game of politics was intense. In the "age of biology," the National Museum had proven successful in its bid to revitalize itself and become an institution that would have a relevant role in defining the nation's new directions. The work of scientists like Roquette-Pinto, Mello Leitão, and Sampaio shaped the museum into a welcoming space for convergent visions of a society guided by biology. It was to this end that they petitioned to be part of the new political journey toward "national reconstruction." And they were able to forge bonds with the Provisional Government because their conception of biology meshed well with the organic, corporatist authoritarianism then holding sway.

A second major factor underpinned relations between the National Museum and the Vargas government. Through a well-organized campaign, the members of the institution were working to transform Brazilian children into well-educated adults, familiar with their homeland's territory and riches, its natural resources, and its fauna and flora. From the moment he took the helm as director, Roquette-Pinto spearheaded a remarkable educational drive inside the National Museum, making creative use of new means of communication. During those years, the museum hosted initiatives that relied on a range of mass media, all against a backdrop of great indifference toward the boundaries between disciplines, even as specialized fields were emerging and the notion of specialization had gained respect. These accomplishments and creative skills did not go unnoticed by the government, and this was another key reason why the National Museum achieved prominence soon after 1930.

2

A MINIATURE OF THE FATHERLAND

It is not hard to understand the great, generalized esteem
in which brasilianos *hold the old institute.*
First of all, it is a miniature of the Fatherland.

—ROQUETTE-PINTO, INTRODUCTION TO *UIÁRA*, 1937

BRAZILIAN BARBECUE

ON NOVEMBER 11, 1930, the Provisional Government—established on October 24 following the victory of the self-proclaimed revolution—gave a barbecue for some of the troops in Rio de Janeiro, then capital of the Federal District. The venue was the gardens of Quinta da Boa Vista, headquarters of the National Museum. Director Roquette-Pinto invited the revolutionaries to visit the museum, where three films were screened in their honor: *Em pleno coração do Brasil* (Deep in the heart of Brazil), *Nos sertões do Brasil* (In the Brazilian hinterlands), and *Carnaúba* (The carnauba wax palm), the latter based on a script by Alberto Sampaio. Five hundred and thirty men, probably in small groups, watched these motion pictures in the exhibit hall that the director had built shortly after he came to office in 1926.[1]

Thirteen days later, on November 24, an even more illustrious guest paid a visit to the museum: Getúlio Vargas himself, "honoring this institute" and "leisurely browsing the collections on exhibit." Roquette-Pinto recounted the event to Minister Francisco Campos in the annual report of the National Museum, which had been attached to the Ministry of Education and Public Health (MESP) following the Revolution of 1930. The director took the opportunity to express his appreciation for having been retained at his post, which he interpreted as a "lofty vote of confidence."[2] Seven months later, in June 1931, both Getúlio Vargas and Francisco Campos attended commemorations of the museum's 114th anniversary, evidence of their continued interest in the institution, which had assumed an important role both in the administration's strategy to strengthen its own legitimacy and also in its project to shape a "new Brazilian man." Vargas and Campos, major strategists of this new era, believed the museum would serve them well in the task of deciphering the complex hieroglyphics represented by the land and people of Brazil. The museum could function as a true "map of legibility" in creating mechanisms of control and homogenization. When they scrutinized the museum's displays, both men were "seeing like a State."[3]

In Roquette-Pinto's opinion, the main reason everyone admired the museum was that it was "a miniature of the Fatherland." Few people would ever be able to travel across all of Brazil, but wandering through the exhibit rooms was like gazing upon "the portrait of a loved one." In a few brief minutes, "the features characteristic of the many regions where our compatriots live, delight, and suffer" unfolded before the visitor's eyes, "in collections within everyone's reach."[4] As Vargas browsed the exhibits in November 1930, was it a similar line of thought that prompted him to place the institution under the umbrella of the newly created MESP? It seems likely. The new government encountered a sophisticated structure for educational activities at the National Museum. The building had undergone extensive remodeling in 1927; three new stories housed a library, the hall for conferences and exhibits, and new workshops (drafting and modeling, photomicrography, typography, mechanics and electricity, book binding, carpentry, and painting). The botanical garden had also been redone.[5]

As argued in chapter 1, the field of biology had been tightening its ties with government since the dawn of the century in Brazil. But this alone does not explain why the Provisional Government valued the National Museum so much. There were other institutes with top-quality researchers, like the Oswaldo Cruz and Butantan Institutes, which had demonstrated their public health policy skills under critical circumstances. What singled out the National Museum at that time was its ability to interweave biology and education. Its scientists were geared to conveying knowledge about Brazil and sharing practices of importance to the nation, not just in the realm of biology—regarding hygiene, flora and fauna, soil, physical anthropology, and nature conservation—but also in the social arena, particularly in educating "the people."

Education was the key for establishing dialogue between National Museum scientists and the Provisional Government; it was the panacea for all troubles, conflicts, and impasses—a "pedagogical illusion." In the mid-1920s, Roquette-Pinto had declared that Brazil would not cure its woes "with either the secret ballot, the organization of political parties, compulsory military service, or a reform of the Constitution." Such "remedies" would only bear fruit if "the general mass of people would be able to vote securely, not shirk their civic duty, obey authority and the law, and work and produce without falling prey to exploitation." As it was, the "people were in no condition to benefit from methods of refinement." No initiative would yield results without "intellectual and moral trailblazing."[6] Two assumptions underlay these thoughts. First, Roquette-Pinto discarded outright the possibility that civic participation or the pursuit of the true exercise of political citizenship could lead to any transformation of Brazilian society in the medium run. Second, in his eyes, an intellectual elite should guide "the people" where they "ought" to go before they could be deemed capable of engaging in political life. Through education, Brazilians dispersed across this huge land would learn to understand and adopt rules of hygiene and scientific teachings and could then do battle with disease, redeem their spirits and bodies, be integrated into the nation, and change their environment. They would, in short, fulfill their destiny.

It was with such expectations that these and other scientists engaged in countless educational initiatives, many with the support of the state and others with the aid of diverse sectors of society. They cast themselves as the guides of a susceptible, ignorant, and childlike people, within whose breast beat a "collective soul," awaiting a summons so that it might reveal itself. In so doing, these men threw their support behind an authoritarian playbook that gradually gained hegemony over the course of the 1930s.

No matter how patent the authoritarianism underlying their attitudes, it is impossible to discount the excitement, passion, and idealism with which these scientists immersed themselves in so many projects. Likewise surprising are their versatility and the ease with which they transited between different fields of learning. Their quest for new media and their bold experimentation with these new languages were hallmarks of their work; they were determined to "remove science from the exclusivist domain of sages and deliver it to the people."[7] After a period of decay that seemed to presage the end of the "age of museums," Roquette-Pinto, from 1926 to 1935, launched a number of initiatives aimed at turning the National Museum into an intellectual center for debate, research, and the construction of knowledge. The goal was to make the museum a disseminating hub of information for society at large through projects for national renewal. The museum brought together scientists from different disciplines to begin interacting as a group organized around specific projects. Some came from Brazil's schools of medicine or engineering, whereas others were self-taught. They had read similar works and were in agreement on many scientific ideas. They had traveled to the interior of Brazil or were familiar with other people's accounts of such journeys. Their visions of the nation's problems and the urgent need for educational action were similar. They belonged to a range of associations in which they constructed social networks, and they had idols like Euclides da Cunha and, in particular, Alberto Torres. The National Museum was a fertile site for a meeting of minds and the staging ground for a multitude of endeavors, both inside and outside its walls, such as the expeditions led by Rondon, Brazil's heated race debate, scientific exchanges, national and international congresses, the creation of radio stations, and Fernando Azevedo's Biblioteca Pedagógica editorial project.

The National Museum was quite adroit at insinuating itself into the process of reaching a new balance of power after 1930. Its staff was extremely creative not only in proposing new ideas but also in organizing attractive new ways to convey this information by conjoining knowledge with intellectual adventure and inventing fun ways to learn. They devised novel educational methods that had the potential to affirm the scientific truths they preached, and they were notable players in the day's political game and its attendant power relations.[8] From their standpoint, to educate was to train workers—good, useful, orderly, wise, patriotic men—but in order to accomplish this task, educators had to be on the same wavelength with people of all ages and be able to appeal to the child within.

These methods were not born in 1930 or invented by the Provisional Government, and they did not simply serve as an "ideological arm" of state power. Rather, they had been evolving for some years as a byproduct of intellectual critiques of Brazil's oligarchies and their disregard for the Brazilian people and its land. From the inception of the republic, there had been debates about a Brazilian style of education, based on revised methods that would kindle a shared sense of patriotism. Examples include José Veríssimo's *A educação nacional* (National education) and Silvio Romero's *A história do Brasil ensinado pela biografia de seus heróis* (The history of Brazil taught through the biographies of its heroes), both released in 1890.[9]

One landmark event was the advent of radio in Brazil. In 1923, Roquette-Pinto assumed a role at the forefront of this enterprise, well ahead of his peers. Through his contagious enthusiasm over this new technology in mass communication, he persuaded a number of colleagues that radio was a viable means of education. He united them around a common project, honed the skills and strategies essential to enlisting the support of public officials, mastered the technical know-how, and coordinated broadcasts with publishing initiatives. Under his influence, scholars from different fields placed their scientific knowledge, time, dedication, political skills, and prestige at the service of the common ideal of education through radio. For Roquette-Pinto, this was a defining moment in his process of acquiring skills that would soon prove essential when he became head of the National Museum.

This chapter begins by exploring the interactions of the group of scientists who experimented with educational radio and acquired a wide range of technical skills, while at the same time working to establish themselves as experts in specific fields, effectively engendering a climate of robust interdisciplinary exchange in which scientific and technological proficiency both contributed. We will see how Rádio Sociedade was a kind of trial run for Roquette-Pinto's grand experiment as director of the National Museum, as he negotiated with political authorities and premiered new media. The chapter next looks at the educational activities conducted by the museum from 1926 to 1935. While projecting itself as an interactive space open to visitor collaboration, the institution also emphasized the transmission of established, prepackaged knowledge—laying bare the authoritarianism inherent in this facet of its educational approach and the ensuing contradiction between this and the touted ideal of participatory transformation of society. The *Revista Nacional de Educação* was one of these projects. An outgrowth of the museum's work in cinema, the magazine was funded by the Film Tax (allocated by the MESP), and in this sense its existence depended on the Commission for the Censorship and Selection of Educational Films, chaired by Roquette-Pinto. The magazine's ultimate demise, in June 1934, reflects the uncertainty of the political moment as well as the precarious status of National Museum staff within the political projects then underway. Lastly, the chapter describes the participation of Roquette-Pinto, Mello Leitão, and Sampaio in the largest publishing project of their time, the Brasiliana Collection, which secured them a definitive place on the intellectual stage in the 1930s.

BROADCASTING

In 1926, the Rádio Sociedade do Rio de Janeiro station, with Roquette-Pinto in charge, launched the magazine *Electron*. The first issue offered an explanation of the English word "broadcasting," a new loanword that had its listeners intrigued. The term, said the magazine, was a compound word combining the verb "cast," a reference to the agricultural act of sowing, and "broad," in the sense of "wide-ranging" or "far-reaching." Ergo, "broadcasting" expressed the action of "sowing into the distance, scattering good seeds far and wide." And since nobody should sow poor quality seeds, "broadcasting should always live up to its name."[10] The notion of sowing meshed well with two ideas that Roquette-Pinto—as the magazine's director and also as secretary of Rádio Sociedade—deemed of paramount importance: farming, which he believed was vital to the achievement of a more authentic sense of nationality, then lying dormant in the unreached countryside, and ethics, embedded in the distinction between good and bad seeds.

The Rádio Sociedade do Rio de Janeiro station was inaugurated in 1923 by members of the Brazilian Academy of Sciences (ABC), which had in turn been founded in 1916 at the initiative of a group of scientists, some of whom were on the staff of the National Museum—for instance, Roquette-Pinto, the zoologist Alípio de Miranda Ribeiro, and the archaeologist Alberto Childe.[11] The academy's mission was to discuss the boundaries between the sciences, cultivate the so-called pure sciences in an impartial quest for truth, organize courses and conferences for the scientific community, bring knowledge to wide sectors of Brazilian society, publish a periodical featuring research findings by its members, and establish scientific awards. At academy meetings, members addressed each other as "scientist," a habit that attested to their desire to form a distinctive identity. They also advocated the demarcation of specialized fields, because they believed that the complexity of each field of science demanded full-time devotion and a depth of knowledge incompatible with the figure of the multipurpose sage versed in the rhetoric of generalist knowledge.[12] Although these scientists spent their academy meetings talking about the need for pure, impartial science and for specialization, they also placed great value on the dissemination of knowledge to society at large. In this crusade to convey science information, they worked side by side on collective projects like radio, absorbing expertise beyond the limits of their specific domains.

In its day-to-day operations, Rádio Sociedade had to solve many practical problems. It became crucial to have both a theoretical and a practical understanding of radio, broadcasting, and equipment and of the challenges of improving reception—the subject of countless articles in the magazines *Radio* and *Electron*. There was nothing irrelevant or boring about it, not at a time when radio had come on the scene as one of the most promising modern peacetime technologies, with an equally powerful potential for times of war.[13]

Roquette-Pinto's curiosity about radio broadcasting was first piqued by his research into physiology. As a medical student, he had attended classes in experimental physics given by Henrique Morize at the Polytechnic School in Rio de Janeiro. Using the lessons of his former professor, he managed to generate a small source of continuous waveforms in the laboratory. In September 1922, when radio was introduced to Brazil at the International Exhibition in Celebration of the Centennial of Independence, he fell in love with its potential as a form of mass media. In his words, Brazil was witnessing "the dawn of radio" and the miracle of "mysterious waves that silently transmit harmony through space." He compared his emotion to those experienced by the men who "owned and read the first books."[14]

After the initial equipment had been installed, Roquette-Pinto persuaded Professor Morize to support the radio project through the ABC. First, however, they had to jump through legal hoops because of the restrictions on these activities. Roquette-Pinto invited Amadeu Amaral, an essayist and member of the academy, to have a look at the radio in hopes of garnering his support. Amaral wrote an article in *O Estado de S. Paulo* describing his surprise at discovering that "this contraption made from bamboo, a few meters of copper wire, a cardboard spool, and an ordinary telephone apparatus" worked wonderfully. In addition, Roquette-Pinto painstakingly researched how other countries regulated radio broadcasting and drew up suggestions for a law on the transmission and reception of radio communications by private parties; the academy then submitted his proposals to the Ministry of Justice and Internal Affairs. Although Rádio Sociedade was not granted its license until August 1923, its first broadcast took place earlier that year, in April, followed by a series of "experimental" broadcasts—and let us not forget that the listeners could have been arrested for possession of *galenas*, as the crude, homemade crystal sets were known. On September 7, 1923—Independence Day—the radio began broadcasting legally, first operating out of the Polytechnic School. In 1924, the station moved to the spacious Czechoslovakian Pavilion, originally built for the Centennial Exhibition, where studios were set up with equipment donated by the Brazilian postal service.[15]

In 1924, Roquette-Pinto and his colleagues at the academy launched the semimonthly popular science magazine *Radio*, a complement to its broadcasts. The publication was replaced in 1926 by *Electron*, headed by Roquette-Pinto and likewise published twice a month. It was no mean editorial challenge. The editors not only had to compile articles and illustrations and then oversee the printing and distribution of the magazine but also to coordinate printed matter with their broadcasts.[16]

The station also had a library, which in 1926 contained eight hundred cataloged books, along with international magazines on radio and scientific journals.[17] The radio had an ongoing campaign to recruit new members, since it depended on membership fees and on advertising revenue from both on-air productions and its magazine. As a membership benefit, the society offered to file the necessary paperwork for members so they could receive a certificate of good standing from the Ministry of Transportation and Public Works, which was a prerequisite for anyone wanting to purchase and install a home radio receiver; the red tape included filing a formal petition and submitting a police-issued certificate of residency. Monday through Saturday, members could also avail themselves of a help desk manned by members of the technical commission, should they need "information on building or repairing radio apparatuses or taking care of defects."[18]

As part of Rádio Sociedade's daily programming, major news stories from Brazil and around the world were read from the daily papers. The schedule also included a children's program, classical music, lectures on a variety of topics, and classes in history, Portuguese, English, French, chemistry, physics, farming, and hygiene.

Rádio Sociedade avowed its independence from any business or industrial interest and its commitment to fostering education and public instruction in Brazil. The station was not supposed to be an end in itself but rather a means for delivering cultural programming to the public. According to Roquette-Pinto, radio broadcasting was the "greatest school of tomorrow," and every home "scattered across the vast land of Brazil" could now receive "the comfort of science and art." In the sertões of Goiás and Mato Grosso and on the arid plateaus of the Northeast, people would now be able to hear music that would enrich, soothe, and refresh the spirits of young and old alike—straight from Rio's opera house. If used "with heart and soul," radio could transform people in a matter of minutes. It was a book for those who did not know how to read, and yet it would also combat illiteracy by awakening within each listener the "irrepressible desire to learn to read." When radio had managed to "land in every backyard on the wings of its infinite flight," compatriots across the country would start working better and producing more. Living in the sertões would no longer condemn a person to "dying alive." The radio would be the illiterate's newspaper, the schoolmaster of those with no school, free entertainment for the poor, and the "spark of new hope, comforter of the ill, guide for the healthy," as long as it was utilized in an "altruistic, lofty spirit."[19]

All this optimism about radio undoubtedly reflected a strong underlying belief in technology as a kind of magic. Roquette-Pinto calculated the number of receivers in Brazil and estimated how many people were reached by each. Imagining people gathered around loudspeakers in every single far-off village, farm, or plantation, he reckoned each receiver was heard by five people, putting at thousands the number of daily listeners who had the benefit of "lessons, lectures, music, the history of Brazil, hygiene, helpful farming tips, news, and information on science." He argued that radio's immense success could be traced to something deeper: the "organic solidarity" of the human species, whose thirst for social relations flames "an unbridled desire to communicate with one's peers." It was also vital to instill certain ethical tenets (the idea of planting good seeds) so that broadcasting stations would not be swept up in "selfish individual interests" but would instead be guided by the greater good.[20]

For all these reasons, Roquette-Pinto pressed the government to subsidize the purchase of radio sets. Once in possession of a modest radio, and even before knowing how to read and write, every single Brazilian—whether he or she be "barefoot or threadbare, ragamuffin, pallid, languid from disease or ignorance"—could learn that "sloth is almost always disease . . . [and] that to be a soldier is not to be a slave but rather to receive instruction and an education, in proper places, led by compatriots fraternally devoted to serving the country."[21]

In addition to being a founding member of the ABC and overseeing the magazine *Electron*, Roquette-Pinto maintained a busy schedule at the National Museum. After being appointed director of the museum in 1926, he resigned from the magazine in August of that year, but he did not abandon his radio activities. One of his first measures as museum director was to try to establish a nationwide educational radio system that would involve state and municipal schools as well as the more educated citizens in communities around the country. After all, every major town has a "worthy Court Judge, [who is] a scholar in the History and Geography of Brazil," a "talented young" attorney devoted to poetry and literature, a physician who could give lessons in natural history and hygiene, women school teachers, young boys who play piano at church, and young girls who sing. These individuals could be recruited "for the sake of educating the poor." A radio station did not cost much, "less than the main altar at the town cathedral." The state should subsidize the purchase of radio sets by making it possible to buy them at cost—and consider these funds well spent on popular education.[22]

The prevailing winds, however, were not that favorable. In 1928, the local administration of the Federal District asked that Rádio Sociedade move out of the Czechoslovakian Pavilion. Roquette-Pinto tried to transfer the station to the National Museum but the costs were prohibitive. Maintaining its independence from the museum, the station leased space elsewhere. With commercial radio stations multiplying, the government issued a decree in 1932 mandating that antennas have a power of at least 5 kW, a standard Rádio Sociedade could not afford to meet. Numerous obstacles notwithstanding, Roquette-Pinto stalwartly led the station until 1936, when it was donated to the government, under the condition that it remain under the aegis of the MESP (in other words, outside the domain of the Press and Propaganda Department) and true to its educational objectives.[23]

Perhaps foreseeing how hard it would be to keep the station on the air, in 1934 Roquette-Pinto joined forces with Anísio Teixeira, then director general of public instruction for the Federal District. Together they founded PRD-5, an educational radio station with modern, powerful equipment, located at the Rio de Janeiro Institute of Education (formerly the Normal School, which was rechristened when Teixeira initiated a thoroughgoing reform, reorganizing it to provide experimental, secular education and undergraduate-level teacher training). The station's programming focused on elementary school courses and teacher training courses. In Teixeira's plans, the radio would play an essential role in extension education at the University of the Federal District, whose mandate included not only teaching but also research and extension. The station was headed by Roquette-Pinto until 1937, when he began devoting himself exclusively to the National Institute of Educational Cinema.[24]

In his speech at the PRD-5 inauguration ceremony, Anísio Teixeira said radio was the medium that would spell the end of the limitations of the past, placing itself at the service of directed education and making up for Brazil's shortage of schools; the microphone was the "classroom for millions of spectators." On the same occasion, Lourenço Filho declared that radio and movies might be sources of corruption but, like fire and water, they could also enrich society by conquering distance, tearing down walls, stealing into all corners, and transmitting "enlightenment and guidance, questions that stimulate and words of comfort." By sending in a request, listeners could receive free course programs, drawings, and maps that would give them a much better understanding of the courses and lectures heard on air.[25]

By working at both Rádio Sociedade and PRD-5, and also as a founding member of the ABC and head of the National Museum, Roquette-Pinto served as a key go-between, as did some of the museum's other scientists. In his broadcasting activities, he assumed leadership of a collective endeavor that spawned manifold offshoots. Alberto Sampaio was one of the most important of the scientists invited to join these efforts, right at the founding of the Rádio Sociedade station. He gave many lectures—on flowering floss-silk trees, proper care of trees, the trees at the Pasteur Institute in Paris, the planting of eucalyptus trees in São Paulo, orchid raising, botany in elementary school and in the Boy Scouts—as well as a complete practical, multi-module course on forestry in 1926.[26] In the early 1930s, when the museum hired Mello Leitão, he was assigned a regular fifteen-minute time slot every Friday on Rádio Sociedade. He lectured on an assortment of topics, including grasshoppers, spiders, sea serpents, genetics, frogs, Cnidaria, eolites and tektites, plant pigments, meteorites, and the genesis of the continents and oceans according to Alfred Wegener.[27]

When Roquette-Pinto took the helm of the National Museum, he launched an ambitious plan to revitalize the institution, no doubt a product of the rich experience he had acquired as leader of Rádio Sociedade. The museum metamorphosed into a promising place for experimentation, new modalities of communication, and new uses of technology, where staff members could team up in a collective effort, obtain government support for projects, and accomplish great things. The museum provided Roquette-Pinto with a sizable team of botanists, zoologists, entomologists, anthropologists, archaeologists, and geologists, complemented by a variety of technicians, during a time of smooth give-and-take between fields (ultimately making everyone a scholar of natural history). Yet it was also a period of increasing specialization, when scientists were aspiring to gain recognition in one specific field. Working together and sharing ideals, expectations, and interpretations of Brazil, they designed their projects in frank dialogue with the political context of their day, blending scientific, technical, artistic, and literary forms of knowledge. This molded an intellectual milieu where different fields interacted, methodologies were exchanged, and gray areas within and between existing disciplines encouraged the development of new knowledge. It was a time of specialization yet in a climate of intense communication among fields. Science, art, and technology all cooperated with each other as part of the bigger nation-building project.

This picture brings to mind what we now call "transdisciplinarity": drawing a team from different fields to focus on a collective project that conjoins science, technology, and art in an environment where borders between the disciplines blur and where sharing experiences is a priority.[28] It might be somewhat anachronistic but not totally wrong to say that the museum presented itself as a transdisciplinary space. And, as we will soon see, it was a multimedia space as well.

THE ASSISTANCE SERVICE FOR THE TEACHING OF NATURAL HISTORY

The Royal Museum was born in 1818 in Rio de Janeiro under an imperial decree issued by Dom João VI. The institution mimicked the model of Europe's encyclopedic museums, which sought to offer a "great universal census" by gathering all they could in one spot and inviting patrons to travel the continents of the globe by walking just a few meters and peering into a few drawers. Species that lived at great distances from each other in space and time could "easily show up among things in one display case and then in the next," pieces in collections whose goal was to represent the world.[29] The idea of founding a natural history museum in Brazil was clearly linked to the exercise of power by the Portuguese Empire, then intent on tightening relations between political power and the natural sciences. As in other European nations, museums and gardens appeared as both agents and products of modern history, "as a space in which ideas about nature, economy, and legitimate authority interacted with concrete policies" of monarchical power.[30]

Major changes came to the institution in the 1870s under the leadership of Ladislau Netto. New fields were consolidated, like paleontology, anthropology, and ethnology. Anyone applying for a position had to take part in a competitive public selection process, and the staff became more professional as a result. Exchange agreements were signed with European, U.S., and Latin American museums. The *Arquivos do Museu Nacional* began publishing research conducted in Brazil, and starting in 1876, courses and lectures were opened to interested parties. However, as part of an imperial, hierarchical, slave-based society, the museum had a very limited target public. According to the *Jornal do Commércio*, attendees at its courses and lectures were ladies from high society, men of letters, civil servants, and—not just occasionally—the emperor himself.

At the dawn of the republic, however, the National Museum declined in prestige as it encountered competition from new institutions. The Paulista Museum and the Pará Museum of Natural History and Ethnography both came on the scene, headed by Hermann Von Ihering and Emílio Goeldi, respectively. Both men wanted their institutes to set the benchmark for excellence and specialization, and they scorned their counterpart in Rio de Janeiro as a "museum of generalities." The establishment in Brazil of experimental laboratories and research institutes concerned with disease control also helped elbow the National Museum into the background.[31] Arthur Neiva, renowned in biomedicine and experimental science, tried to enforce a project to make the museum more dynamic during his tenure as director, from 1923 to 1926. However, he got caught up in other ventures, particularly the fight against the coffee berry borer and his effort to create the Biological Institute of Agricultural and Animal Defense, so the museum saw no major changes.

Roquette-Pinto had witnessed these troubles, since he had been on the museum staff since 1906. Twenty years after joining the institution, he accepted the post of director with the firm intention of implementing new practices that would authoritatively secure the museum its rightful place on Brazil's scientific and intellectual stage. The anthropologist had just returned from a trip to the United States, where he had visited the American Museum of Natural History, then one of the main hubs of a broad movement to endorse biological studies as a resource for forming young citizens and for national renewal in the United States. The American museum sponsored courses and guided tours, coordinated activities with public schools, and worked in partnership with teachers, professors, and intellectuals from the fields of biology, philosophy, and education, adopting the practices of the "new museum movement" that had been blooming in the United States since the late nineteenth century. This transformation was part of the U.S. context, where museums were being taken over by emerging groups of biologists, eager to foster the public, educational tasks of these institutions. The chief engineer behind these new museum practices was George B. Goode, who said that "a thorough education and knowledge of science and art are vital to the nation and to the place it holds at present in the civilized world."[32] This euphoric, optimistic climate of belief in the transformative power of a museum that produces knowledge while working closely with schools and society at large—all as part of a project of national renewal—dovetailed with Roquette-Pinto's hopes for Brazil. His visit to the United States reinforced his educational convictions. Soon after he returned to Rio, he was appointed director and threw himself into a ten-year period of tireless work.

Roquette-Pinto's administration was guided by the precept that a museum should not be a "mere treasure trove of collections" or just a "center for research into high science, be it in the laboratory or in the field." These were crucial tasks, but in addition the "Institute" (the term Roquette-Pinto used whenever he wrote about the museum) should also devote itself to public education "through all means within its grasp."[33] By sticking to this path, the National Museum would leave its past behind, moving beyond the amassing of natural history collections, and it would set itself apart from other leading research institutions of its day, like the Oswaldo Cruz Institute in Rio and the Bacteriological Institute in São Paulo. Roquette-Pinto sought to carve out a unique niche for the museum at the national level. He opened its doors and exhibit halls to visitors every Tuesday through Sunday, mornings and afternoons. Cleaning services came in on Mondays, while its scientists and other staff carried on with their usual activities.

In October 1927, Roquette-Pinto inaugurated the museum's Assistance Service for the Teaching of Natural History. In his mind, it was imperative to establish laboratories and experimental centers for middle school students. "It's time for us to train researchers," he wrote. Brazilian research institutes lacked skilled personnel. There was no need for extravagant investments; modest laboratories and a good library would suffice.[34] The Assistance Service was an independent department within the museum and was headed by Roquette-Pinto until August 1935, when staff member Paulo Roquette-Pinto—Edgard's son—stepped into the job.

The Assistance Service helped steer and facilitate the organization of "school museums," whose holdings included native specimens that had been collected and prepared by the students themselves under the guidance of their teachers. There were courses and lectures on how to capture insects and small mammals and how to gather plants and minerals. Participating schools sent materials to the museum, where they were prepared, mounted, and classified in the Assistance Service room. The lists of material that was received illustrate the variety of things sent to the museum, including samples of wood, seeds, insects, stones, shells, eggs, and bones. Attached to the 1929 report, a photograph of the room shows boards of butterflies, stuffed animals, shells, and books, as well as bottles filled with all sorts of substances, assembled in a hands-on environment where technology, science, and education cooperated with each other (figure 5).[35]

The Assistance Service also prepared educational guides, charts, slide shows, and posters for use in schools, along with publications in popular science. During the years that followed, teachers, students of all ages, and Boy Scouts packed

FIGURE 5. Room where specimens were prepared, mounted, and classified by the Assistance Service for the Teaching of Natural History, 1929. SAE 146.5, Doc. 7.09A, folder 105. Courtesy of SEMEAR.

the museum rooms for film screenings, slide shows, lectures, and courses of all kinds. Classes focused on practical museology, taxidermy, histological techniques, the collection of animals, the organization of herbaria, photography (lighting, picture taking, developing, and printing), drawing, and wax molding. Given the museum's stagnant budget, a great deal of creativity was needed to set up and maintain equipment. For example, a photomicrographic camera that used natural lighting was fashioned out of "improvised, makeshift equipment on a rough wooden stand, with a small discarded 9×12 camera" (figures 6, 7, and 8).[36]

Although the full name of the Assistance Service referred only to "natural history"—which was the term then used to designate the class taught at schools—the underlying epistemological perspective encompassed the teaching of biology in its broadest sense. The goal was to instill in students the habit of observing relations of interdependence in nature so they could "better understand our own life," in healthy contrast to older natural history methods.

FIGURE 6. Workshop of the Assistance Service for the Teaching of Natural History, 1929. SAE 146.5, Doc. 7.09A, folder 105. Courtesy of SEMEAR.

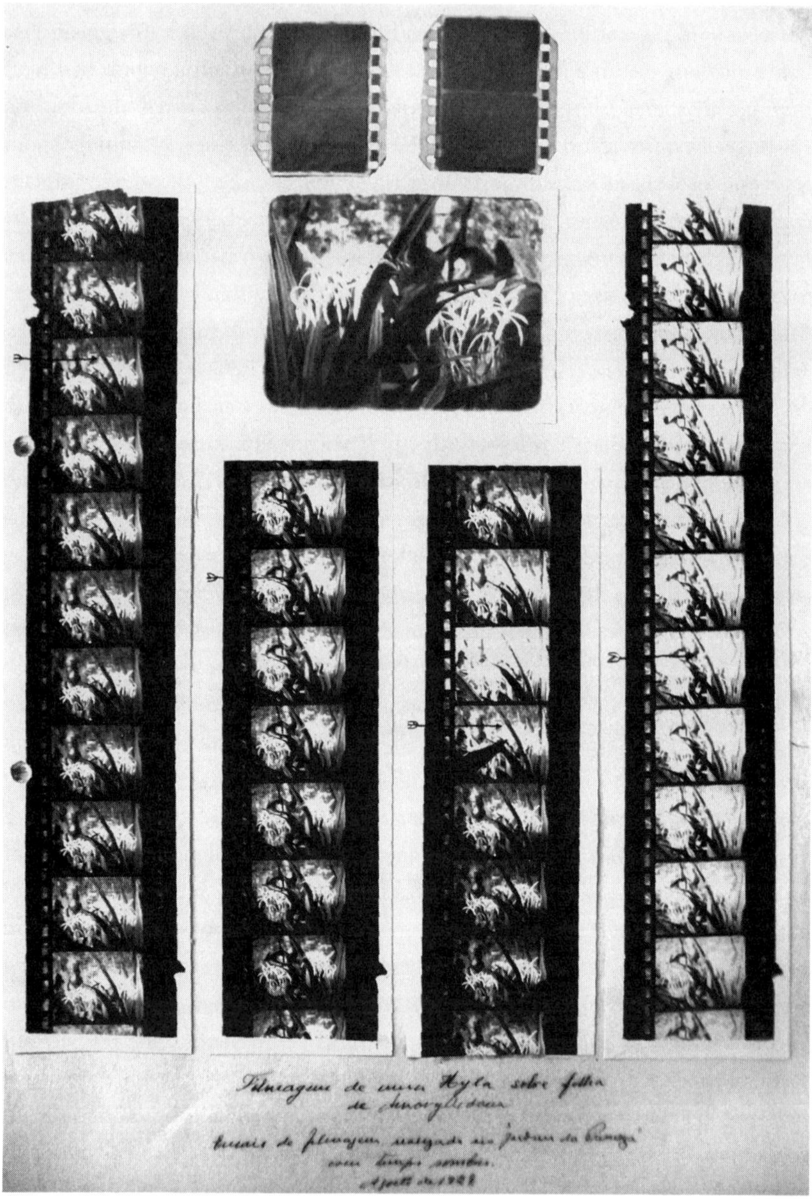

FIGURE 7. Negatives for educational films, 1929. SAE 146.5, Doc. 7.09A, folder 105. Courtesy of SEMEAR.

FIGURE 8. Print shop at the National Museum, 1929. SAE 146.5, Doc. 7.09A, folder 105. Courtesy of SEMEAR.

Instead of memorizing systems and names, students would take up "the inquisitive study of life, delving into true biology."[37]

The Assistance Service also worked with the University of Rio de Janeiro, founded in 1920; it made use of the college's facilities and technical equipment and offered extension courses under its umbrella. In 1932, for example, professors from the National Museum gave the following courses in partnership with the university, all of which included weekly theoretical and practical classes, slide shows, and films: Popular Biology (Roquette-Pinto), Spectral Analysis Techniques Used in Mineralogy (Alberto Betim Paes Leme), Phytogeography (Sampaio), Scorpions and Other Small Venomous Arachnids of Brazil (Mello Leitão), and National Studies in Brazilian Ethnography (Heloísa Alberto Torres).[38]

Sampaio, who had been with the museum since 1904, shared the director's enthusiasm. He pictured museums in every school in the near future, their work coordinated with scientific institutes and especially the National Museum. This would make it possible to track the distribution of species and their frequency in different botanical and zoological zones in order to arrive at a detailed survey, something that would not have been feasible earlier, since there had been no way to ascertain "what actually exists in each region." These school museums would not only generate local knowledge but also enhance the production of knowledge at research centers.

Sampaio exemplifies how the National Museum networked with schools and associations at various levels in a process of mutual reinforcement. The botanist gave countless lectures at agricultural clubs in schools throughout the interior, led annual botany contests at schools, was active in lending guidance to the Friends of Nature Clubs created at municipal schools in Rio de Janeiro, and was a founding member of both the Society of the Friends of Trees (1931) and the Society of the Friends of Alberto Torres (1932). He promoted lectures on nature, botany, and forestry at the Rotary Club and the Touring Club do Brazil (an association to attract foreign tourists and to foster tourism among the Brazilian elites as well). He worked with teachers from myriad city and country schools to hold commemorations related to Arbor Day and to gardening and the planting of seedlings. He tied all these activities in with the museum, which acted as a source of knowledge and guidance. In his work at schools and with women's associations, Sampaio had the assistance of the botanist Bertha Lutz, a museum staff member since 1919, who had graduated from the Sorbonne in 1918 and was a prominent feminist.[39]

WHAT IS A MUSEUM'S PURPOSE?

During one of his lectures on Rádio Sociedade, Mello Leitão, who had worked as a zoologist in the museum's invertebrates department since 1931, spoke about the "Educational Role of the National Museum of Natural History." According to the professor, a false notion of museums still held sway: they continued to be viewed as repositories for rarities and for bizarre plants and animals never seen before. But, said Mello Leitão, a museum's mission was actually much different; it should teach and display the "most authentic, truest facets" of nature. Generally speaking, ninety-nine out of a hundred Brazilians who had visited the galleries of a museum knew absolutely nothing about the real lives of the organisms in its collections or about their characteristics or habitats. Museums had to be maintained precisely with these people in mind—"and that was exactly how things should be."[40]

In order to be the "repository of everything that exists in the country," the National Museum relied on the dedication of its scientists. But this was not enough. So the museum called on the general public to collect small animals in the region where they lived and send them to the institution; the specimens should be the most common, the most ordinary, and the least surprising, like bugs, centipedes, snails, scorpions, and spiders. The museum gave detailed instructions on how to carry out the task: how to protect oneself from any danger the animals might present, how to choose the right glass jar and seal it hermetically, how to use alcohol to preserve specimens, how to package the specimens safely, and how to mail them.

Mello Leitão saw this as a two-way street: as a member of the museum, he wanted to share the knowledge being produced there with the rest of society, but he also wanted his audience to assist the institution with an educational project, in which they would become actual collectors for the museum. The exhibits should showcase "what is typically Brazilian," exploring the structure, morphology, and ecology of gathered plants and animals so that "our culture" becomes increasingly "more familiar and more cherished."

This movement could have an even more dynamic impact: once material had been received from collaborators nationwide, it could be transformed into illustrations and prints for publication in textbooks or the *Revista Nacional de Educação*, forming a "veritable iconography of Brazilian zoology" and thus disseminating to all regions of the country information on plants and animals from distant areas as well as those common across the land. Teachers and professors at schools scattered throughout the nation could then teach biology without having to resort to "the exotic figures found in European books." Studies would be more dynamic, thanks to the readily identifiable images that portrayed plants and animals well known to students.

The goal of awakening a new vision of plant and animal life in Brazil might also have been a response to vestiges of the age-old controversy about the value of the plants and animals that flourished in the New World, and specifically in Brazil. In the mid-eighteenth century, the Comte de Buffon and Cornelius de Pauw argued that living creatures in the Americas were inferior and that nature had degenerative effects in these lands. Such ideas ignited a great polemic. Thomas Jefferson not only contested Buffon in the pages of his *Notes on the State of Virginia* but also went to great lengths to send him a giant taxidermied moose as concrete proof of the magnificence of American wildlife. In the nineteenth century, Hegel was one of the leading champions of the idea of American degeneration. Thinkers like Humboldt and Darwin contributed substantially to quashing these ideas, and slowly the polemic died out. To some extent, the museum's overall posture was a rejoinder to the echoes of this debate; it defended the excellence of the animals and plants found in Brazil, worthy of the keenest attention on the part of scientists and of other people as well. In other words, the museum validated the merit and grandeur of Brazil's flora and fauna.[41]

The museum's overriding goal was to accumulate a multitude of references that could be shared, because knowledge of nature stood as a powerful means of instilling love of nation as part of an educational project that connected the National Museum, its publications, and its exhibits with schools all over Brazil. Also featuring prominently was an acute appreciation for experimental knowledge, grounded in real life and in everyday experience. The very meaning ascribed to the physical space of the science museum and to its collections underwent a revision.

Roquette-Pinto thought of the museum as a strategic staging ground for educational action tailored to youth. The exhibit halls allowed young people direct contact with nature, encouraged them to get to know their own environment, and replaced a natural history of wonders with a natural history of ordinary things out of everyday life—a prerequisite for strengthening their love of their homeland. In Roquette-Pinto's opinion, children were like the Brazilian people in miniature and displayed fetishistic, "wild souls." They could not be expected to understand their country through abstract notions but only through "the shaded orange grove and the brook where tadpoles swim about in dark shadows of schools . . . [and] the land itself, with its scrublands and its birds, its beaches, sand, and sea." They needed to learn about commonplace, pedestrian things and vanquish their ignorance of the names of plants and animals. In the interior, the use of popular names for living creatures reflected this dearth of knowledge, for every animal there was "just . . . a critter."[42]

It was not enough for the museum to organize exhibits that flaunted the rich diversity of Brazilian nature. Roquette-Pinto insisted that exhibits should leave an indelible mark on the patron's soul. Otherwise, the visitor would stroll apathetically by the display cases, "like a drop of water rolling off a greasy slide," without being moved or transformed by anything. A visit to the museum should be a powerful experience, absorbing visitors deeply in time and space. They should linger there awhile, observing objects, attending lectures, looking at posters, and watching movies and slides—the last two both highly innovative resources back then. When they entered the building nestled in Quinta da Boa

FIGURE 9. Boy Scouts on a visit to the National Museum, December 1927. Série Documental Museu Nacional. Courtesy of Arquivo Múcio Leão, Academia Brasileira de Letras.

Vista, children and their teachers would experience something unique; they would enjoy a different kind of interlude from their regular school or family life, for they would discover an environment filled with special objects, and the museum scientists were sure they would be mesmerized by them. The scholars that assembled the displays wanted people to make the very most of their time during what might be their one chance to visit the museum. When the scientists welcomed these guests or when they just walked around the museum sporting lab coats, they were presenting themselves as laboratory researchers engaged in experimental science and nourishing the image of the museum as a research center (figure 9).

Learning was to be facilitated by the use of tantalizing forms of media. Any excitement caused by an exhibit was not an end in itself but rather a strategy; everything was designed to ensure an effective educational experience. The exhibits themselves were forms of media, designed for the purpose of communicating. The intent was also to spark a collective experience; the children walked through the museum and filled its spaces in groups, aided by guides and

FIGURE 10. Teachers leading students on a class trip to the National Museum, 1930s. *Revista de Educação Pública* 8:33–40 (1951–52): 51. Courtesy of Fundação Biblioteca Nacional.

teachers. The goal was to breed a sense of belonging to the nation, to kindle an identity through the activities that had been devised collectively by the museum staff, and to mold the young visitors into Brazilians—or *brasilianos,* as Roquette-Pinto liked to say, rejecting the much more common *brasileiros*. In the hallways and galleries of the National Museum, Brazil was transformed into a showy spectacle (figures 10 and 11).[43]

Photographs of exhibits and patrons at the museum show shelving units laid out in orderly rows, arranged so that lines of students could file by, stopping to stare at objects through transparent glass. In one of these photos, the students hold pencils and paper to record their observations. They are looking at material gathered by the traveling naturalists hired by the museum in the second half of the nineteenth century, as well as at collections of specimens brought back by the scientists who accompanied Rondon. Roquette-Pinto himself organized the Hinterland Ethnography collection, selecting the utensils, instruments, and materials to be included in the display.[44] There was a sharp awareness of the paths these objects had traveled before reaching the museum and how they had been chosen and arranged alongside other objects. But another important consideration was how they would be viewed and appreciated by visitors.

FIGURE 11. Anthropology exhibit hall, National Museum, 1929. SAE 146.5, Doc. 7.09A, folder 105. Courtesy of SEMEAR.

Films and slides were shown in the spacious Marajó Hall. Inaugurated in 1932, it offered comfortable chairs and a décor inspired by Marajoara pottery. There, according to an article from the magazine *Cinearte*, "a marvelous Krupp Ernemann apparatus" showed the 150 films "carefully stored in cans and numbered." The idea was to optimize visitors' comprehension of the objects on display at the museum: after looking at "stationary" shelves, they would witness an "animated exhibit"; after looking at a "cold and lifeless octopus," they would watch "moving" images, thanks to the magic of film. "A modern museum without cinema is not a museum," the director pronounced. Motion pictures transformed a visit to the museum into a voyage through space and time, freeing visitors from the fate of "living and dying cloistered inside the walls of their customs and their era."[45] The museum also put great store in the impact of overlapping media: display cases, slide shows, posters, lectures, guidebooks—and all of this infused learning with "an iterative renewal."

The museum was designed as a space for interaction and iteration. Patrons were meant to lend a hand in constructing knowledge while at the same time new means of communication and technologies would be used to teach them established knowledge through repetition, relying on an assorted pallet of tones, colors, images, and sounds for this purpose.

When Mello Leitão suggested that the public collect material and do the initial preparation so it could be shipped to the museum for exhibition or to illustrate magazines and textbooks, he was envisioning listeners, readers, and visitors taking up the active job of being "voluntary collectors." As such, they would view with new eyes the "critters" in their region, now imbued with a certain "aura" and a whole new meaning—after all, these animals had deserved a scientist's attention and were worthy of being displayed in a museum. Anyone would find it wondrous, stimulating, and rewarding to see "their spiders" or "their critters" duly labeled in Latin and shown off in a display case or transformed into drawings that illustrated the *Revista Nacional de Educação* or the pages of a textbook. People would have a fresh new attitude toward the animals and insects in the places where they lived.[46]

When Sampaio brought teachers, students, and members of diverse associations together to organize school museums and gardens, he too was interested in active involvement in the production of knowledge. The same notion lay behind Roquette-Pinto's proposal for municipal radios. In other words, the National Museum's introduction of interactivity was a pivotal step in its project of renewal. These interactions would be bolstered by experimental teaching about everyday life, local landscapes, and immediate, palpable challenges—teaching that at the same time would demonstrate that building knowledge was a collective activity.

Still, the endeavor remained true to the rote learning approach, as Roquette-Pinto revealed when he used the telling expression "iterative renewal" in his interview. Content was transmitted via sound waves, in the form of classroom radio programs, the news, and handpicked music. The same content appeared in print format in magazines, which featured summaries or full transcripts of the museum's courses and lectures. This content could also be seen in the moving images shown in Marajó Hall, in slide shows, and on the printed posters and charts distributed to schools or included in museum guidebooks. On the one hand, this rendered each and every object on display more dynamic, since it became part of an array of networks; on the other, since everything was served up on a platter of explanations, the result was a passive learning experience that deprived visitors of the opportunity for true active participation. The logic of repetition was grounded in the "pedagogical illusion" of a childlike people who were like blank sheets of paper or lumps of unmolded clay, needing only to discover what their masters already knew. The emphasis on collective experience was also part of the plan to guide the people. This pedagogical perspective was thus quintessentially paradoxical: while efforts centered on enabling people to construct knowledge themselves, the prevailing assumption was that they would reach foregone conclusions. This reinforced the perspective of a republic from the top down, with an intellectual crème de la crème at the fore.[47] And so an authoritarian political culture was marched out yet again and gained new momentum in the early 1930s with the support of the Vargas Provisional Government and the creation of the MESP. No matter how innovative its methods, the National Museum's project for renewal was still conservative in its ends.

IN EVERY HOME

In October 1932, the hallways and exhibit rooms of the National Museum were bubbling over with excitement. The first issue of the *Revista Nacional de Educação* (*RNE*) was about to roll off the presses, and hopes were high that all 12,500 copies would enjoy wide distribution across Brazil, a Herculean task assigned to the Directorate of Information, Statistics, and Dissemination, an MESP department. Teachers, schools, mayors' offices, and cultural and professional associations began receiving free subscriptions.[48] The magazine was a key piece in the National Museum's broader crusade to renew Brazilian society.

A lucky recipient of the first issue would see on the cover the image of a stately yet delicate woman reaching up to touch a radiant source of light.[49] Was this a depiction of Minerva, the protective goddess of all intellectual activities and particularly of schools? If so, missing were her usual staff and shield—possibly a providential oversight for a magazine published by a government that touted social pacification. Did she denote the Brazilian nation? The luminous rays shone across a somewhat obscure landscape, where we can make out the lines of houses and mountains in a setting perhaps urban, perhaps rural. On the right side is the title of the magazine and the name "Ministry of Education and Public Health," followed by the tagline that would be stamped on every issue: ". . . in every home in Brazil, the moral comfort of Science and Art" (figure 12).

The *RNE* traced its existence to the Instructions to Decree 21.240, of April 1932, which nationalized the film censorship service and instated a "Film Tax for popular education," which was levied on the exhibition of movies and calculated on a per meter basis. The Provisional Government felt that motion pictures—"a form of entertainment the public can no longer do without"—had great potential in the realm of popular culture as long as they were "duly regulated." The tax was intended to fund the following: the establishment and maintenance of a National Institute of Educational Cinema, the importation of unexposed film for the production of educational documentaries, a film library at the National Museum, and, last but not least, the publication "of a popular magazine in science, literature, and arts communication" to be distributed to all public teaching institutions. While the *RNE* was a direct expression of the educational concerns and initiatives of the MESP, its history also had much to do with the government's newfound attention to cinema, viewed both as a threat and as a potential educational tool in building and strengthening an idealized nation.

The same decree also established a Censorship Commission for films, comprising a representative of the chief of police, someone from the Juvenile Court, the director of the National Museum, a teacher, and a representative of the Brazilian Education Association (ABE). Roquette-Pinto was chair of the hard-working commission, which reviewed 1,200 films in its first year alone; some were classified as educational, others had scenes deleted or were banned in their entirety, and still others were classified as inappropriate for children or minors (see figures 13 and 14). Not long afterward, the government handed down new

FIGURE 12. Cover of the first issue of *Revista Nacional de Educação*, 1932. Courtesy of Sistema de Bibliotecas da UFMG.

instructions that were more standardized and detailed and that established a Commission for the Censorship and Selection of Educational Films; these instructions were signed by Anísio Teixeira, then head of the Institute of Education. It is important to note that all of these actions were centralized within the MESP; furthermore, in practice, the commission concerned itself more with educational matters than with purely moral issues—in fact, it was sometimes criticized for its excessive lenience with regard to the latter.[50]

Printed on plain paper, the monthly magazine measured 7 by 10.5 inches and averaged ninety-six pages. Light and compact, the publication's physical simplicity was no doubt intentional, with form complementing content. Once its pages were opened, however, the magazine proved impressive. It offered about sixteen

FIGURE 13. "Fox Film salutes the *Revista Nacional de Educação* and presents an exclusively educational motion picture, the first and only shot entirely in the heart of the African desert! The secret of the African jungles unveiled by the derring-do of men! Watch a white woman do battle with an enraged rhinoceros! Watch a fight to the death between two gorillas! *CONGORILLA!* With Mr. and Mrs. Martin Johnson. This documentary film was two long years in the making! Special Brazilian edition, with explanations in Portuguese." *Congorilla* poster, *Revista Nacional de Educação* 1:1 (1932). Courtesy of Sistema de Bibliotecas da UFMG.

articles per issue, most short and always written in purposely straightforward, educational language. The magazine was supposed to be inviting—the more accessible it was, the more effective as a tool.

Contributors to the *RNE* were distinguished scientists and intellectuals. A good share of them were on the staff of the National Museum, like Roquette-Pinto, Mello Leitão, and Sampaio, as well as Alberto Childe, Carlos Vianna Freire, Moysés Gikovate, and Raimundo Lopes. The magazine published excerpts by major authors as well, especially Alberto Torres but also great names

FIGURE 14. "Magic Carpet: The world before your eyes in images and sound! Real motion pictures that entertain and *educate*!" *Tapete Mágico* poster, *Revista Nacional de Educação* 1:3 (1932). Courtesy of Sistema de Bibliotecas da UFMG.

like Euclides da Cunha. Some public officials made their way into its pages too, in the form of transcriptions of their speeches on educational matters. Educators, teachers, professors, and members of the Society of the Friends of Alberto Torres also contributed sporadically. Likewise noteworthy were articles on Brazil by eminent naturalists, some of which were rare texts. The twenty-one issues published from October 1932 to June 1934 offered sixteen excerpts translated directly from Spix and Martius's *Reise in Brasilien (Viagem ao Brasil)* and six fragments of *Viagem filosófica* (Philosophical voyage), by the Portuguese naturalist Alexandre Rodrigues Ferreira.[51]

In the quest to win over its audience, one of the magazine's most successful features—besides its pantheon of authors—was its extensive use of high-quality black-and-white images of National Museum holdings or reproductions of classic works of art from around the world. These illustrations fulfilled a number of purposes. For one thing, they made the magazine more attractive. There was no effort to catch the reader's eye with the cover, which—save for the first issue—was habitually quite solemn. On the left side, there was always a small sketch of the bust of a countryman in whom Brazilians could take pride, like Alberto Torres, Euclides da Cunha, Nísia Floresta, Pedro Américo, José Bonifácio, Carlos Gomes, Santos Dumont, or Diogo Feijó (figure 15). But the unpretentiousness of the cover stood in stark contrast to the images and beautiful photographs inside, which illustrated many of the articles and were printed on better paper than that used for text pages. The message to the reader was that the periodical's outer simplicity belied the profound content concealed inside, since the *RNE* had made an aesthetic choice to reject any extravagant or misleading appearances. Stripped of what Alberto Torres called the "glitter of discourse," the magazine wanted to distance itself from the values of so-called coastal culture, whose vanity and ostentatiousness were seen as breeding affectation. Only when the reader opened the magazine and carefully perused its pages would its secrets be divulged. This was true even of the cover, because its deeper meaning could only be found in the Notes and Information section, which offered a brief biography of the person being honored each month, highlighting his published works and other contributions to Brazilian society.

One of the photographs showed an indigenous man from Rondônia—taken by none other than Roquette-Pinto himself. Other photos portrayed illustrious Brazilians, like the painter Vitor Meirelles and the inventor Santos Dumont. The magazine also featured photographic reproductions of classic paintings by masters like Rembrandt, Da Vinci, and Michelangelo, along with works by

FIGURE 15. Cover of *Revista Nacional de Educação* 2:6 (1933).
Courtesy of Sistema de Bibliotecas da UFMG.

Brazilian artists from the National Academy of Art or Itamaraty Palace, such as *Caipiras negaceando* (Hinterlanders on an ambush), by Almeida Junior, and *Bandeirantes* (Frontier adventurers), by Henrique Bernadelli. Biographic information and the titles of other works by the same artists could be found in Notes and Information; in the case of Brazilian artists, exhibits and awards were also listed. The fact that Brazilian artists were intermingled with their foreign counterparts and received just as much fanfare signaled their worth. Another goal was to foster aesthetic appreciation as the cultural foundation of the new Brazilian man. The magazine wanted to provide what it called "the moral comfort" of both science and art, firmly coupling moral elevation with the building of a national culture and sensitivity, in tune with the educational trends of the day.[52]

The *RNE* also printed long excerpts from the travel accounts of Alexandre Rodrigues Ferreira, together with some of his previously unpublished drawings, such as *Índio Cambeba atirando flecha com a palheta* (Cambeba Indian shooting an arrow with a spear thrower), as well as other images of his: an armadillo against a landscape that had obviously felt the human hand, a snake and a capybara, a capuchin monkey, and a hut of Cururu indigenes. These reproductions appealed to the reader's aesthetic sensibilities while conveying knowledge that was invaluable to a society that had learned about Rondon's journeys and was still meeting the challenge of integrating the Amazon into the nation (figure 16).

Some images were placed seemingly at random, without any relation to a specific article—for instance, depictions of Brazilian landscapes like the Tijuca forest, Iguaçu Falls, and the city of Ouro Preto. Many others were strategically placed to enhance comprehension of a given text. Biology texts were illustrated with images of spiders, ants, leaves, stems, and trees, while texts on hygiene were accompanied by drawings of the human body (figure 17).

In short, between its covers, the magazine delivered a plethora of information. Simple and unassuming on the outside, it disclosed a whole world to its readers—not a strange and far-off world but a world that would be built into the nation of Brazil. Harmonizing form and content, text and image, the *RNE* clearly intended to construct a shared set of references and address them to a well-educated public comprising teachers, professors, and cultural associations, hopefully reaching peer educators in their midst. Published and distributed free by the Provisional Government, the *RNE* was most assuredly a strategic weapon used by the administration to validate its new political project—a project wherein the scientists of the National Museum consciously strove to negotiate a new role for their institution. The government allocated the magazine an

FIGURE 16. Print from *Viagem filosófica*, by Alexandre Rodrigues Ferreira. *Revista Nacional de Educação* 1:7 (1933). Courtesy of Sistema de Bibliotecas da UFMG.

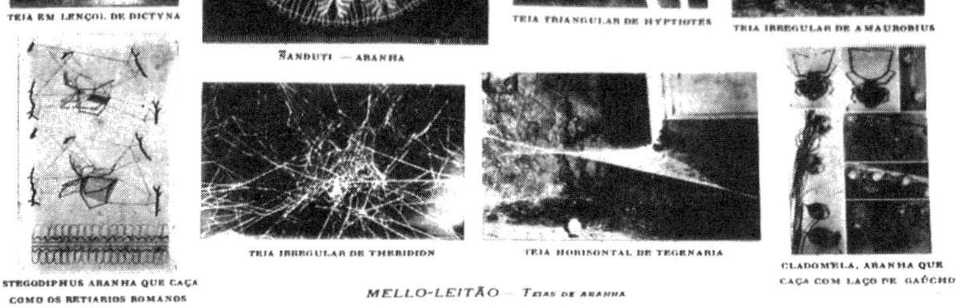

FIGURE 17. "Mello Leitão—Spider Webs." *Revista Nacional de Educação* 2:9 (1933). Courtesy of Sistema de Bibliotecas da UFMG.

annual budget of some eight contos de réis. Factoring in advertising revenue and outlays on material, plates, editing, and other labor, each issue was estimated to cost five hundred réis, at a time when the price of a daily newspaper was two hundred réis.[53]

Brazil's sheer size and its precarious transportation system were obvious hindrances to circulation, but Roquette-Pinto deemed this problem simply part of the battle to be won; modern conveniences and ideas crept into Brazil slowly, and the very vastness of the territory to be conquered indicated the grandeur of the work under construction. The determination to carry the magazine to far-flung, forsaken corners of Brazil as a herald of state initiative reinforced one of the many ideas then being discussed by intellectuals, that is, that the sertões lay at a great distance from public power and from the government's modernizing projects. And yet—as the renowned writer Afrânio Peixoto had famously declared in 1922—Brazil's sertões also lay right on the nation's doorstep, just a little beyond the end of Avenida Central, then a major thoroughfare in downtown Rio de Janeiro, capital of the republic.[54]

The illustrations in the magazine were a fundamental contribution to the defense of these arguments because they constructed an image of Brazil. The inside covers regularly featured information meant to signify a general idea of the Brazilian nation and its history, people, and territory. In the case of issues 4 to 10, the inside front cover always featured a map entitled "Brazil and Its Borders," delineating the country's boundaries with its Latin American neighbors while simultaneously giving a notion of the vastness of the territory to be settled and civilized (figure 18). The inside back cover for these issues displayed a table of statistics on Brazil in 1930. Many in Brazilian society were then calling for data like this, including champions of statistics like Fernando Azevedo, director general of public instruction for the Federal District from 1927 to 1929, who organized the first school census in Rio de Janeiro; another was Teixeira de Freitas, who urged Brazil to get to know itself "in order to become the master of its fate," saying that the country knew almost nothing about itself or "the splendid legacy that divine benevolence had bequeathed it."[55] This information was meant to be a source of inspiration for Brazilians, exhorting them to value their natural heritage and their potential as a nation.

Issues 11 to 15 featured other images on their inside covers. The inside front covers showed a map of population zones, drawn up by Roquette-Pinto. These maps underpinned his argument distinguishing "race" from "people"; in his view, the cultural and historical aspects of the category "people" made it the preferable

O BRASIL E SUAS FRONTEIRAS

1 — Foz do Oiapoc. 2 — Serra Tumuc-umac. 3 — Rio Tacutú. 4 — Rio Maú. 5 — Serra Parima. 6 — Cucui. 7 — Rio Uaupés. 8 — Tabatinga. 9 — Rio Javarí. 10 — R. Purús e seus afluentes, S. Rosa e Chambuíaco. 11 — R. Acre. 12 — R. Madeira e seu formador Mamoré. 13 — R. Guaporé. 14 — Morro Quatro Irmãos. 15 — Baía Negra (Rio Paraguai). 16 — Rio Apa. 17 — Serra Maracajú. 18 — Rio Iguassú. 19 — Rio Uruguai. 20 — Rio Quaraim e coxilha de S. Ana.

Os nomes sublinhados são os principais ex-contestados obtidos pelo Brasil.

P. — Pirará — ex-contestado obtido pela Guiana Inglesa.

Nomes em maiúsculas verticais — Republicas independentes.

Nomes em maiúsculas inclinadas — Colonias.

Nomes em maiúsculas menores — Estados e Território Federal do Acre.

FIGURE 18. "Brazil and Its Borders." *Revista Nacional de Educação* 1:7 (1933). Courtesy of Sistema de Bibliotecas da UFMG.

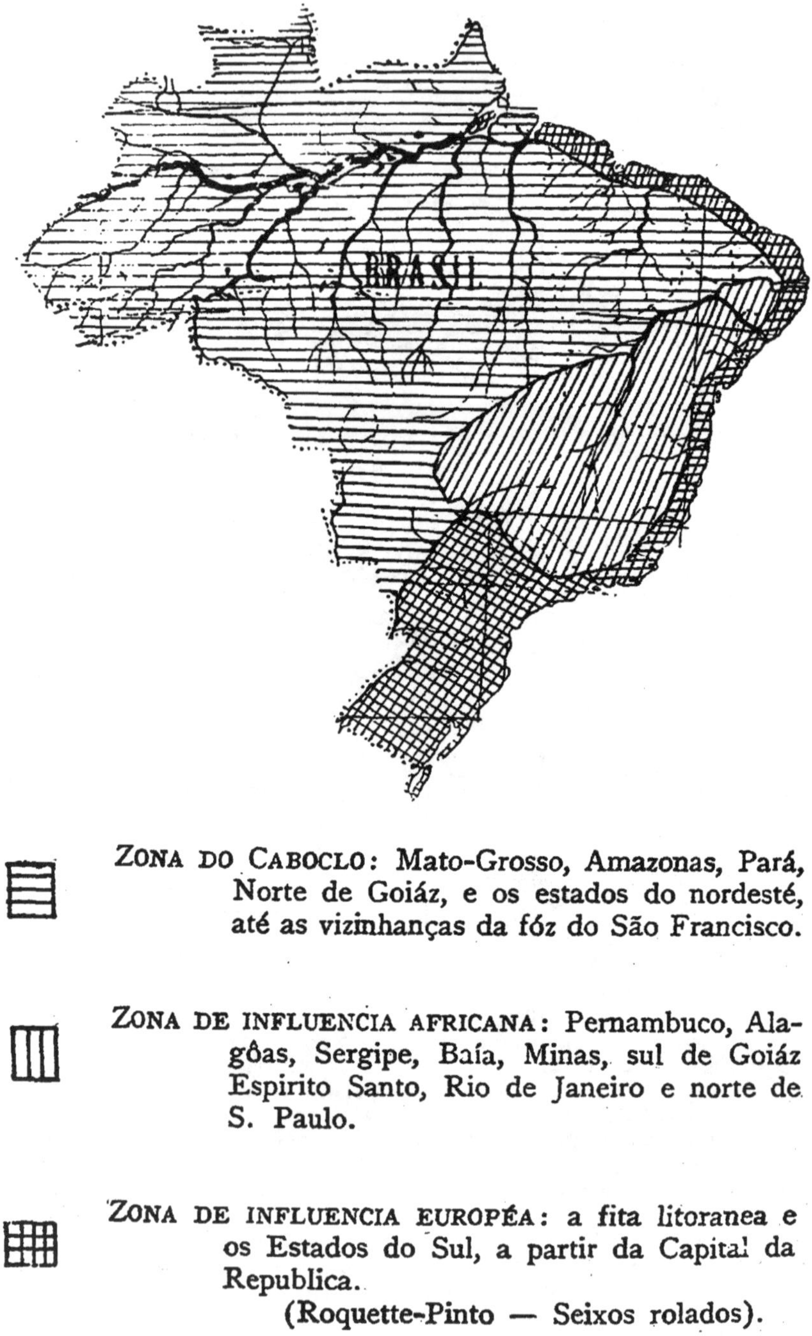

FIGURE 19. Map of population zones by Roquette-Pinto. *Revista Nacional de Educação* 2:15 (1933). Courtesy of Sistema de Bibliotecas da UFMG.

term. Distinguishing between the zones of *caboclos* (people of combined Amerindian and European descent), those of African influence, and those of European influence, the map was a product of the author's thinking as laid out in his article "Brazil and Anthropogeography" (in Portuguese), published in 1927 as part of Roquette-Pinto's collection of essays on Brazil entitled *Seixos rolados* (Rolling stones) (figure 19).

The inside back covers highlighted certain key dates in Brazilian history, starting with the country's discovery and then moving on through the French and Dutch invasions, the private expeditions into the interior of Brazil known as the *bandeiras*, independence, the founding of the Brazilian Historical and Geographical Institute, the construction of the railroad linking the cities of Rio de Janeiro and Petropolis, abolition, the Proclamation of the Republic, the work of Oswaldo Cruz, the Rondon Commission, the institution of the Provisional Government under President Vargas, finally culminating in an event that seemed to lend meaning and hope to this long journey: the establishment of the MESP.

One inside cover of a later issue featured a geological map of Brazil, related to an article on the topic by the National Museum's Moysés Gikovate. Other inside covers showed images of the Greek alphabet and Morse code—the latter relevant because of the monumental importance of Rondon's journeys to the building of the Brazilian telegraph system in distant reaches of the country. Over its last six issues, the inside covers of the magazine featured a world map tracing the routes of the major voyages of discovery. The very last issue also displayed busts of the great navigators, along with brief biographies. These images offered the reader a range of interpretations about Brazil's territory, population, natural and physical characteristics, economy, and, particularly, its history, all from a perspective that blended people and territory, society and nature, to form a representation of the nation.

THE ABCS OF LIFE

When the *RNE* compiled these words and images as vehicles of knowledge and then published them in the hopes that they would reach ever wider audiences, its strategy was to present itself as a kind of school primer that was varied, diverse, versatile, and grounded in hard facts, unveiling a world of signs to be decoded, as if Brazil were a giant book to be read and people needed to be properly equipped for the task. The articles, many of which were part of sections continued across several issues, imparted elementary ideas from different fields of science, the humanities, and the arts.

For one thing, the magazine wanted to be a primer on how to interpret Brazilian flora and fauna, a subject in which the National Museum had a time-honored tradition of study and in which its researchers had played a significant role. A number of contributors worked at the museum, like Carlos Vianna Freire, who appeared in every single issue, in the section "Elementary Notions of Botany." Written in simple language and interspersed with an abundance of explanatory drawings, the texts in this section provided a thoroughgoing course on leaves, stems, and roots. Another constant presence was Sampaio, who wrote about basic principles of botany, the history of plant taxonomy, and methods of plant classification. Some of his other pieces reflected his involvement with the National Museum's film library (which received a portion of the funds allocated under the same decree that launched the magazine). Writing about babassu and carnauba, their phytogeography, and their uses and economic potential, Sampaio explained that his texts were the notes for educational movies to be produced by the National Museum.[56]

Much as the *RNE* promised to decipher the mysteries of Brazilian flora, it also suggested that it would unveil the secrets of its fauna. Various authors wrote articles on wasps and ants, always emphasizing that these were social insects, "beings of the multitude," whose collective behavior was analogous to "the spirit of the multitude that characterizes human society."[57] Mello Leitão's contributions were among the most plentiful and highly diversified. He regularly drew from elements of daily, commonplace life, using myth, legend, and short stories to create an engaging educational experience—a goal that took precedence over specifically scientific considerations.[58]

Sowing knowledge in every way possible, teaching the "masses" to read "the great book of Nature," without which they would never learn "to understand the ABCs of life"—this was "the most indispensable of literacy training and the hardest of all." Such was the goal of the section "Agricultural Matters," written by Otto Frensel, which taught rural men proper farm practices, like how to get rid of mosquitoes and worms, how best to prune trees, and how to observe eggs, larvae, and cocoons in order to tell the difference between insects that were pests and those that were beneficial to crops. The weather could also be an object of curiosity and a source of knowledge. The section "Meteorological Dictionary," written by Joaquim de Sampaio Ferraz, director of the Meteorological

Institute, taught the basics of climatology with entries in an A to Z format. The author covered essential practices, tools, and ideas, which were clearly useful in a nation-building project that was experiencing a growing appreciation for the countryside and agriculture.[59]

A section entitled "The Brazilian Sky" was published from January through December of 1933. It reproduced star charts made by the astronomer Louis Cruls for the Cruls Mission, one of whose key members had been Henrique Morize, mentor and friend of Roquette-Pinto at Rádio Sociedade. Originally published in 1896 under the title *Atlas celeste* (Celestial atlas), the charts showed the firmament at the latitude of Rio de Janeiro as visible on the fifteenth of each month. In the introduction to the series, the *RNE* pointed out that these guides would be very useful for engineers, geographers, and explorers by helping them identify the stars and thus determine geographic coordinates. Although the charts were of the heavens over Rio, they could be used anywhere in the country by making minor adjustments to take latitude into account.[60] So in addition to being a primer on plants, animals, agricultural techniques, and climate, the magazine also encouraged people all over the country to survey the heavens and find their spot in the vast territory that was Brazil.

The section headed by Othelo Reis related mathematics to daily life, an approach aligned with the Escola Nova. Concepts like proportion and infinity were explained, as were the rules for adding long series of numbers, taking odd measurements, doing sums quickly, and calculating time of day at different latitudes as an aid to travelers. Some of these columns were written by Professor Jonathas Serrano, who used math to give a better understanding of history—for example, how to figure out what century a year falls in, how to understand calendars, and what historical ages are all about.[61]

The archaeologist Alberto Childe, founding member of the ABC and one of the magazine's most steadfast contributors, wrote about ancient cultures, especially the Greeks, Romans, and Egyptians. Did this contradict the proposal of proffering knowledge about Brazil? Not at all, because there was a larger goal behind the lengthy explanations of ancient vanities, Greek vases, Roman mirrors, and the reading of hieroglyphics (the author taught readers how to write "National Museum" and "Quinta da Boa Vista" in Ancient Egyptian).[62] After all, how else could museum patrons understand the multitude of objects in the exhibit halls devoted to Greece and Rome or in the majestic hall displaying Egyptian mummies and sarcophagi? How else to lend meaning to their visits? Childe was doing nothing more than teaching the magazine's readers how to read the museum, arming them to better peruse its halls and enhancing the educational power of its exhibits.

The magazine also showcased the museum's new trends in anthropological research, particularly through articles on anthropogeography by Raimundo Lopes. Lopes believed that the debate about the budding science of human geography had only just begun in Brazil, through contact with the work of authors like Carl Ritter, Élisée Reclus, and above all Friedrich Ratzel. But the *RNE* also featured writings by Euclides da Cunha, who imparted a general view of the Brazilian land and people, and by Alberto Torres, with his in-depth exploration of the natural sources of social life. More recently, Lopes explained, there was the work of Roquette-Pinto—for example, his ethnography of the rural poor and his studies of Brazil's *sambaqui* shell mounds and *estearias* (the remains of dwelling sites)—and articles by Heloísa Alberto Torres on the geographic spirit of Marajoara ceramics.[63]

Another series of articles was likewise intended to teach readers different interpretations of Brazil. Moysés Gikovate approached the task in a very eclectic fashion, addressing Brazilian geology, the origin of Brazilian legends, and Brazilian literature and its authors, styles, and eras. He also wrote about the origin of calendars and gave illustrated explanations of how primitive peoples made fire.

Keeping pace with these teachings on how to read the nation came suggestions on how to write it, in sections like "Drawing Lessons," signed by a contributor identified only as Seth, and "Lectures on Photography," by Guerra Duval. Drawing was considered a tremendous aid to the production of knowledge, one that facilitated its systematization and dissemination, exemplified by the generous use of illustrations in the pages of the *RNE*. Roquette-Pinto always contended that drawing could be a central ally to education because it could hold students' attention—of special concern given the "impatient, undisciplined" nature of most Brazilians.[64] The articles taught perspective, shading, and other artistic techniques. In the realm of photography, different types of cameras and methods for developing pictures were presented. Like drawing, photography was cited as a friend to science, mainly because it allowed movement to be visualized through film, but also because it made it possible for the microscopic world to be recorded. Furthermore, artworks that previously could not be replicated were now accessible to the public in the form of faithful reproductions, as featured in the magazine. Lastly, the *RNE* showed that photography was blossoming into an independent art form, where the photographer's work bore the marks of personal interpretation.[65] Because these sections of the

RNE wanted to prove that drawing and photography were within everyone's grasp, they urged readers to do their own sketches and take their own snapshots of landscapes, loved ones, and plants and animals, a task that demanded powers of observation on top of artistic creation.

In this regard, the *RNE* was also a guidebook for training the aesthetic sensibilities thought to be characteristic of culturally superior people, capable of partaking in the construction of a great nation. Alongside popular science articles and reproductions of great artworks, readers found poems and even musical scores, in a country where the piano had been a relatively common instrument since the mid-nineteenth century. These scores were simplified renditions of patriotic Brazilian songs, like the Independence Anthem, the Anthem to the Flag, and the Anthem of the Proclamation of the Republic. Published along with their lyrics, this music taught readers how to play and sing the nation's most important songs.

And what about those who knew nothing of music and for whom scores were unfathomable codes? The magazine had a section just for them: "How to Listen to Music," written by Friar Pedro Sinzig at the special invitation of Roquette-Pinto. In order to "provide the key" so that readers could "step into this immense palace of 'One Thousand and One Nights,'" the magazine gave them a guided introduction to the universe of music. In order to really enjoy music, they had to learn its language, elements, words, and grammar. Friar Sinzig wrote about sound, the timber of different instruments, kinds of interpretation, and the meaning of terms like staccato, legato, pianissimo, and andante.[66]

As we can see, the *RNE* was busy in many different areas, communicating science and art; coaching its readers to get to know Brazil through the country's flora and fauna and its physical, social, and historical features; and instructing them as well on how to find pleasure in works of literature, the fine arts, and music. It also sought to transform every reader into a collaborator in the project to educate the Brazilian people and shape the nation. A number of issues contained samples of literacy posters, which could be utilized by anyone willing to do battle against illiteracy, perhaps motivated by the challenge posed at the bottom of the poster in figure 20: "Are you Brazilian? Are you a foreign friend of Brazil? Don't abandon the illiterate to their fate. Work for Brazil!"[67]

The *RNE* also publicized revolutionary literacy methods, like *Utilinda brincando* (Utilinda has fun), a technique that had been submitted to the MESP by a public school teacher. Through educational games and flash cards that connected phonemes with familiar images, an older, literate child could use this method to teach a number of younger ones to read, and the learning experience

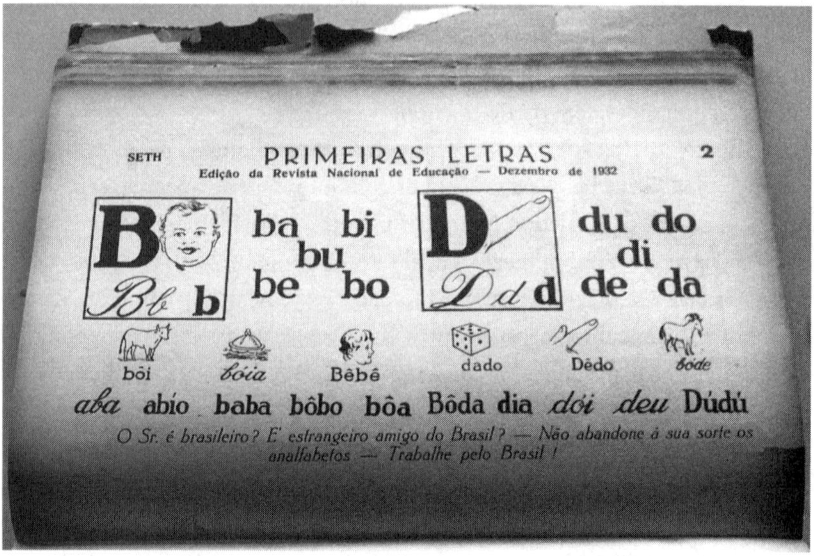
SETH PRIMEIRAS LETRAS 2
Edição da Revista Nacional de Educação — Dezembro de 1932
B b ba bi bu be bo
D d du do di de da
bôi bóia Bêbê dado Dêdo bode
aba abío baba bôbo bôa Bôda dia dói deu Dúdú
O Sr. é brasileiro? E' estrangeiro amigo do Brasil? — Não abandone á sua sorte os analfabetos — Trabalhe pelo Brasil!

FIGURE 20. Literacy lesson. *Revista Nacional de Educação* 1:3 (1932). Courtesy of Sistema de Bibliotecas da UFMG.

could be part of enjoyable extracurricular activities or everyday situations. The magazine featured an article by the same teacher, complete with detailed information and illustrations, along with an enthusiastic appeal to its audience. The name of the character Utilinda was a play on words that combined the Portuguese adjective *útil*, or "useful," with the adverb *ainda*, meaning "still" or "yet"—an allusion to the method's strong suit: it was useful yet fun. The magazine risked an optimistic prediction: if it managed to print 15,000 copies from its first anniversary on, and if each one of these were distributed to schools and associations and reached readers who were willing collaborators, then one child trained to be an Utilinda could teach ten other small children to read and write—and soon thousands of young Brazilians would benefit (a figure just as hyperbolic as the estimates of the reach of radio).[68] Here again, we have the image of a people to be "shaped" by educational initiatives.

Printed matter, sound waves, moving images—in these men's dreams, education should avail itself of every kind of mass media in order to overcome distances, integrate Brazil's huge territory, and deliver people from their reputed ignorance, all part of the task of "explaining the nation to itself."[69] Working in tandem with the institution's areas of radio and cinema, its Assistance Service for the Teaching of Natural History, and its laboratories, workshops, and print shops, the *RNE* was one of the strategic initiatives meant to revitalize the role of the National Museum.

BRASILIANA

Roquette-Pinto rejected the term *brasileiro* to describe the nationality of his country's citizens. He wanted to adopt a new term that would induce every last man and woman born in Brazil to assume a change in attitude, outlook, and practice. Like the suffix "-er" in English, the Portuguese suffix "*-eiro*" often denotes a profession. Brasileiro would thus be a harvester of brazilwood, harking back to the exploitation of colonial Brazil by those who came to carry off the country's resources and make a living from a destructive extractivist economic activity. At the dawning of a new era, as Roquette-Pinto saw it, someone who had been born in Brazil could only be designated by a Portuguese suffix that denotes nationality—and thus could only be a *brasiliano* (much like the English "-an" or "-ian," as in American or Brazilian).[70]

Brasiliana (a term parallel to "Americana") was the title that Fernando Azevedo chose for a series of books in the Biblioteca Pedagógica (Educational Library), headed by Azevedo himself and launched in 1931 by the Companhia Editora Nacional publishing house. The Brasiliana Collection set out to be "the greatest work of Brazilian nationalist culture," and its tagline gave voice to Roquette-Pinto's thinking: "Unveiling Brazil to Brazilians, making it ever more familiar so that it will be ever more beloved." There is no way to separate the history of this editorial project from "a political and intellectual agenda to refound the nation," which sought to nationalize culture and mold "the political face of the country in the 1930s." We can also discern here the ideas of Monteiro Lobato, first owner of the Companhia Editora Nacional, who once said that "a Nation is made of men and books." In the assessment of Minister Gustavo Capanema, the Brasiliana Collection eventually came to embody these ideals as "a living portrait of Brazil." In its first ten years of life, Brasiliana published two hundred titles, a truly remarkable number for the time. The collection set about offering information on a succession of topics and "amassing knowledge and information that open[ed] the way for reading Brazil." Pointing out "paths to a rewriting of Brazil," it stood as a "metaphor of the nation."[71]

The Brasiliana Collection was part of the bigger Biblioteca Pedagógica, which comprised five series: Scientific Training, Textbooks, Children's Literature, Education Today, and Brasiliana.[72] The solidification and expansion of the publishing market, urbanization and industrialization, the growth of the middle class, public teaching reforms, the establishment of new colleges and other institutions of higher learning (including Brazil's first universities), and the artistic and cultural excitement of the day formed a backdrop vital to creating more readers and to this "emphasis on the study and understanding of national reality."[73] More than meeting existing demand, the project was all about blending educational concepts with editorial strategies in order to "invent" readers. By standardizing the format, paper, and covers of the books in the collection, as well as unifying distribution and advertising plans, publication costs were lowered. The five series in Biblioteca Pedagógica were directed at different audiences. Children were the public for both Children's Literature and Textbooks; teachers, for Education Today; and nonstudents, for Scientific Training. Brasiliana was tailored to an adult audience with a sound intellectual background and was divided into anthropology and demography, archaeology and prehistory, biography, botany and zoology, correspondence, law, economics, education and teaching, essays, ethnology, philology, folklore, geography, geology, history, medicine and hygiene, politics, and travel.[74]

Roquette-Pinto, Alberto Sampaio, and Mello Leitão were in high demand as authors during the first decade of the Brasiliana Collection, a sign of their stature in the day's intellectual circles and in the work of constructing knowledge about the nation. The diversity of the classifications into which their works fell within the major divisions of the collection also speaks to the enormous interdisciplinarity of their texts.

ROQUETTE-PINTO'S ANTHROPOLOGY AND SAMPAIO'S BIOGEOGRAPHY

Roquette-Pinto wrote three books for Brasiliana. Released in 1933, *Ensaios de antropologia brasiliana* (Essays in brasiliano anthropology) was listed in the collection's catalog under the headings Anthropology and Demography. It contained sixteen texts related to these subjects, including population and nation, race and education, Brazilian anthropological "types," hygiene and eugenics, heritability, trends in eugenics and biology and theoreticians in these fields, and the First Brazilian Eugenics Congress, in 1929. The book also explored more wide-ranging themes, like the thinking of Alberto Torres and his construction of a national organism, Brazil's shortage of labor power and internal migration, debates on Japanese immigration, and the "brasiliano dialect." Roquette-Pinto engaged in a dialogue with prominent works and authors from the field of biology, including Charles Davenport, Francis Galton, and H. S. Jennings. The book included a bibliography, a list of the main topics explored in each essay, and, unusual for the time, an alphabetical index that allowed readers to locate authors, concepts, or topics.

In 1935, a third edition of Roquette-Pinto's *Rondônia* was published as part of the Brasiliana Collection, in the category of Ethnology. Recounting the author's participation on a Rondon expedition in 1912, the book was plentifully illustrated with photographs, maps, and drawings; it also had lists of indigenous words, phonograms of songs, and an alphabetical index of authors. A fourth edition was released in 1938, this time with footnotes written by friends of the author, including Alberto Sampaio, Mello Leitão, Raimundo Lopes, and Heloísa Alberto Torres, all from the National Museum. Sampaio and Mello Leitão carefully edited the scientific names of plants and animals in the book.

Illustrated and divided into three parts, the last of Roquette-Pinto's contributions to the collection—*Ensaios brasilianos* (Brasiliano essays)—was released in 1940 in the category of Essays. Part 1 focused on some of the personalities admired by Roquette-Pinto: Fritz Müller, Friar Leandro do Sacramento, Alberto Torres, Henrique Morize, Emília Snethlage, Manoel Bonfim, and Miguel Couto, among others. Part 2 delved into themes from Euclides da Cunha's classic *Rebellion in the Backlands*, frontier adventurers (*bandeirantes*), pure science in Brazil, settlement, and racism. In the third part, the author saluted the work of Afonso de Taunay and Miguel Osorio.

Sampaio also wrote three books for Brasiliana, all packed with illustrations and all listed in the category Geography—even though each book's opening page introduced the author as a professor of botany with the National Museum. The first book, *Phytogeographia do Brasil* (Phytogeography of Brazil), was the offshoot of a course held at the National Museum in 1932 in partnership with the University of Rio de Janeiro. Divided into two parts, the first discusses Brazil's floral heritage, introduces the concept and theories of botanical geography, and then explores Brazil's different plant zones. The second part deals with research and covers bibliographic, methodological, and conceptual debates in phytogeography. The theme of protecting nature is the common thread that runs through the two sections. In addition to presenting a conceptual and methodological discussion of the discipline, the book also provides much information and many suggestions on environmental public policy. Endorsing the rational use of nature, Sampaio suggested the issue of forestry as a fulcrum for changes in people's relationship with nature in Brazil and argued that this would require the state to approve strict forestry laws. The government should promote forestry as well, so the country's woodlands would become a source of wealth and valued as such. Sampaio also called for reforestation and the creation of natural reserves under the guardianship of the state.[75]

The second book, *Biogeographia dynamica* (Dynamic biogeography), released in 1935, came with a foreword by Roquette-Pinto, who lauded Sampaio's solid experience: Sampaio, he says, had visited some of the world's leading herbaria and scientific institutions, described newly discovered plants, and traversed "leagues of woodlands, fields, and scrubland." Through his books, wrote Roquette-Pinto, Sampaio wanted to lay the scientific groundwork for the protection of nature and for phytogeography in Brazil, proposing the latter as a new field. Moreover, the sociologist could not be separated from the naturalist, so "thoroughly entwined are [his] proclivities." Inspired by Alberto Torres, Sampaio should be read "in all homes and all schools."[76]

In this particular book, Sampaio offers his observations on the protection of nature in different countries around the world; he provides information on international debates, a chronology of conferences on the topic, advances in legislation, interventions by public power, and initiatives by scientific, educational, and civil societies. He also begins compiling a list of Brazil's natural monuments, sketches out some ideas for tourism, gives some suggestions for the government, argues in favor of more country schools and new reforestation laws and measures, and pinpoints specific regions that he feels warrant official protection.

Published at a strategic time, the book was a centerpiece in contemporary debates. The previous year, Sampaio had served as rapporteur-general at the First Brazilian Congress for the Protection of Nature, sponsored by the Society of the Friends of Trees on April 8–15, 1934. In 1935, the National Museum published a report on the event in two issues of its journal, *Boletim do Museu Nacional*. The congress had been sponsored by Getúlio Vargas, and a good number of public authorities attended the opening session. All science sessions were held at the National Museum under the chairmanship of Roquette-Pinto. Other museum members took an active part as well, including Mello Leitão, Carlos Vianna Freire, Heloísa Alberto Torres, Raimundo Lopes, and Moysés Gikovate. Several associations were represented there too, like the Society of the Friends of Alberto Torres, the Touring Club do Brasil, the ABE, the ABC, and the Brazilian Historical and Geographical Institute.[77]

The weeklong event included poetry readings and musical presentations and showcased educational activities that valorized Brazilian nature, like the Friends of Nature Clubs at Schools. There were assessments of legislative advances in the protection of nature and recommendations for reforestation initiatives. Papers and conferences were presented at eight sessions: soil and subsoil (caves, underground rivers, mines, geological monuments); flora (plants, reforestation, forestry); fauna (rare animals, hunting, fishing, entomology, birds); anthropology (children, eugenics, rural habitats, the protection of indigenes); sites, landscapes, and monuments (scouting, tourism, roads, architecture); education (rural education, eugenic education, instruction in the protection of nature); and legislation and laws. There was also an accompanying exhibit of movies, radio conferences, maps, and art.[78]

During the congress, Roquette-Pinto and Sampaio submitted a proposal to create a national park that would be an especially rich resource for the study of marine biology and tropical flora and fauna, extending from Guanabara Bay to the peak of the Serra dos Órgãos Mountains outside Rio de Janeiro. Under their proposal, the mountains of this Atlantic Forest region would be reforested with Amazonian species; the National Museum would manage the park and also set up a few laboratories for scientific research on its grounds. Sampaio also proposed that a number of parks be established in the vicinity of different towns for the enjoyment of Boy Scouts. Roquette-Pinto expressed his regret that nothing was really being done along these lines and pointed out that his proposals had found resonance among different sectors of society. Along with the Touring Club do Brasil and other associations, he said, a "chorus" could be heard "urgently clamoring for brasilianos to quit being men who make or enlarge deserts."[79]

It was a critical time for museum scientists, who were pushing for public policies to protect nature. The Forest Code, the Game and Fish Code, the Law on Scientific Expeditions, the Water Code, and the Mine Code were all decreed in 1934. Programs and reports from the Congress for the Protection of Nature are rife with flattering references to Vargas and repeated thanks to the government for its sponsorship—no doubt a strategy for maintaining and even boosting this support.[80] Sampaio's book was written and published in the heat of these events and undoubtedly aimed at reinforcing what he felt was a favorable political climate. He wanted to establish a new field of scientific knowledge at the junction of biogeography and social thought, one that would validate the idea of strong, centralizing measures to protect the environment and advance efforts to get the government to do more. As the author proudly said, "The landscape has been cleared for the protection of nature in Brazil."[81] Above all, he expected that scientists at the National Museum would be in the forefront of drafting and enforcing nature laws and regulations. In the report mentioned earlier, Sampaio stated that the "masses" should be guided by a small number of thinkers. From his point of view, this actually paid tribute to the people, "pliant to good teachings."[82]

The main theme of Sampaio's last book, *A alimentação sertaneja e do interior do Amazonia* (Eating habits in the sertões and the interior of the Amazon), is also the question of guiding the "docile masses." The book contains echoes of Alberto Torres in its insistence on encouraging people to remain in the countryside, valuing rural Brazil, and fostering ways to ensure prosperity and permanent settlement there. Defining the sertões as an "uncultured, underpopulated land," Sampaio outlines strategies for securing regional self-sufficiency by leveraging biogeographical knowledge. He evaluates food resources and examines the potential for changing the eating habits of the rural poor by combating taboos, introducing new crops, and promoting certain dietary and hygienic practices. All of this is pivotal, the author argues, at a moment when public power is concerned with "integrating the nation's life forces." The principal form of action would be to organize a "unit of the Boy Scouts in the sertões," which, "through the effect of scouting uniforms, graduation ceremonies, marches, and excursions," would entice hinterlanders to embrace the principles being advocated."[83]

The second part of the book contains a detailed glossary of the food, beverages, and spices consumed in rural areas, as well as observations about the food preferred by the more well-to-do living in the interior. It includes a list of harmful, inebriating, and toxic plants and has a vast bibliography and brief alphabetical index. Describing the book as a continuation of *Biogeographia dynamica*, Sampaio said he found inspiration in the thinking of Pereira Barreto, Alberto Torres, Oliveira Viana, Alberto Rangel, Cândido Rondon, Roquette-Pinto, Arthur Neiva, Belisário Penna, and Getúlio Vargas, among others.[84]

THE BRAZILIAN FABRE

Mello Leitão was a tireless author for the Biblioteca Pedagógica; he wrote five titles for the Brasiliana Collection, two for the Scientific Training series, and two compendiums for the Textbooks series. His first Brasiliana title, *Visitantes do Primeiro Império* (Visitors to the First Empire), is richly illustrated and has a foreword by Afonso d'Escragnolle Taunay, then director of the Paulista Museum and a member of the Brazilian Academy of Letters. Taunay compares Mello Leitão to Roman Emperor Septimius Severus, "who murmured the most noble of advice as he took his last breath: *laboremus semper*!" (labor always), and presents him as an "arachnologist of renown and the author of major scientific publications and widely used textbooks." In this book, Mello Leitão collates countless excerpts from the writings of circumnavigators who were in Brazil between 1815 and 1840. In so doing, he composes a narrative of the coastal cities that sheltered the travelers between legs of their journeys—ports like Recife, Salvador, and Rio de Janeiro. The book is based on works that remain largely unread today, works the author intended to excavate "from the dust of libraries inside museums and institutes" and make familiar to "those who love Brazil and are interested in its life"—words in perfect harmony with the goals of the Brasiliana Collection. According to Mello Leitão, the period in question was one of the most important in the history of Brazil, which made access to the knowledge held in these travelers' accounts especially valuable.[85]

In 1937, Mello Leitão published three books: *Zoogeografia do Brasil* (Zoogeography of Brazil), *O Brasil visto pelos ingleses* (Brazil as seen by the British), and *A biologia no Brasil* (Biology in Brazil). *Zoogeografia do Brasil*, another illustrated work, was no doubt intended as a reference book. It has a lengthy alphabetical index of cited authors and a detailed index of fauna, with both common and scientific names. In his introduction, the author emphasizes the merit of the subject matter in an age of "promising awakening to what is ours." The book came out after Sampaio's *Phytogeographia do Brasil* and claimed to use the same analytical approach, with the goal of crafting "a harmonious, uniform vision of the whole, representative of the National Museum's thought on Brazilian Biogeography."[86]

O Brasil visto pelos ingleses was also in the line of works meant to disseminate information from travelers' accounts of the country, in this case the British. It was a less ambitious book than *Visitantes do Primeiro Império*, lacking illustrations, index, and foreword. It had been nearly 130 years since the Portuguese court had moved to Brazil, bringing waves of travelers in its wake, including a substantial number of Brits—merchants, adventurers, naturalists, tourists, engineers. The narratives describe aspects of Brazilian nature but also scenes from everyday life, customs, and political events. Many of the works used by Mello Leitão were hard to come by and had not been translated.

A biologia no Brasil has a foreword by Roquette-Pinto, who characterized the volume as a reliable guide to the history of biology in Brazil, indispensable for anyone who wants to explore the topic and written by a learned, "top-notch naturalist," master of a robust, clear, and graceful style.[87] The table of contents corroborates the author's intention to demarcate the field of biology in Brazil by framing it within the nation's past; it is not a book of natural history but dares to lay claim to the term "biology." Yet it does not hint at any rupture between the two fields; rather, it suggests a link between them, as if—taking a linear, cumulative view of knowledge—natural history had been a kind of "childhood" of biology. With chapters on each century in the history of Brazil, the book traces the roots of the field back to the 1500s. The analysis then moves on to contemporary Brazil, with special chapters on the state of research in zoology, botany, anthropology, anatomy, and physiology. The book closes with a detailed alphabetical index of cited authors, once again laying bare the author's and publisher's shared intention of creating a reference work.[88]

Mello Leitão's last title for Brasiliana was *História das expedições científicas no Brasil* (A history of scientific expeditions in Brazil), an outgrowth of his work as rapporteur for the Congress on the History of Brazil, held in Rio de Janeiro in 1938. The book is devoted entirely to scientific expeditions that sought to survey and study nature. Its analysis is divided into two parts: "A terra" (The land) discusses accounts of the discovery and exploration of Brazil's coastline, borders, rivers, soil, and riches, while "A vida" (Life) presents botanical, zoological, and ethnological findings from the expeditions. There are no illustrations, although the book does have an index of cited authors.

In addition to writing these five titles, Mello Leitão collated, translated, and annotated a collection of texts about the Amazon, written by the Dominican Gaspar de Carvajal in the sixteenth century and by the Jesuits Alonso Rojas and Cristobal Acuña in the seventeenth century.[89] Mello Leitão also translated and annotated (with over 500 footnotes) *The Naturalist on the River Amazons* by Henry Bates, published in Portuguese under the title *O naturalista no Rio Amazonas*. He stated that the book should "be read and reflected on by all Brazilians," since it suggested that there was "nothing more agreeable, nothing more empathetic than our people and our culture."[90]

Mello Leitão was one of the few authors who had a hand in other Biblioteca Pedagógica series, in addition to his five titles for Brasiliana. Under the grouping Textbooks, he published a four-volume basic course on natural history and a single-volume work on general biology (*Curso elementar de história natural* and

Biologia geral).[91] He also wrote two best sellers for the series Scientific Training: *A vida maravilhosa dos animais* (The wonderful life of animals) and *A vida na selva* (Life in the jungle).

A vida maravilhosa dos animais was released in 1935 to great critical acclaim and was reviewed in a number of newspaper literary columns. Múcio Leão, of the Brazilian Academy of Letters, wrote a glowing review in which he said Mello Leitão had the ability to combine the rigor of science with the pleasure of learning intriguing facts about the animal world. Lúcia Miguel Pereira, a reputed writer of short stories, emphasized that the book had prompted her to ponder how the study of animals could impart lessons on human life and society. Another, an anonymous appraisal, stated that the book was a voyage "through the enchanted kingdom of zoology" which entertained as much as it taught; there was much to praise and much to gain from this "book by the master," according to this unnamed reviewer, because its explanation of the "complex and harmonious" machinery of life revealed the *prima causa*—which rules, foresees, and provides all things. Maurício de Lacerda, a member of the National Liberation Alliance, pointed out how the descriptions of parasitism and commensalism by Mello Leitão—whom he called "the Brazilian Fabre"—shed light on various practices of the Brazilian elites, an interpretation no doubt unauthorized by the conservative, religious Mello Leitão.[92]

Lacerda was referring to Jean-Henri Fabre (1823–1915), the French entomologist famous for his opposition to Darwin's theory of natural selection. Fabre tended to indulge in flights of poetry, the key to the popularity of his books. He was well received among Europe's conservative and Catholic groups because of his dogmatic apologetics for the humility of intelligence before the unknowable and his perception of biology as a true and natural catechism of theology.[93] The comparison with Fabre was far from gratuitous—the cover of *A vida maravilhosa dos animais* featured a large photograph of the French entomologist (figure 21).

Mello Leitão's book was a compilation of lectures that interpreted nature from an eminently political perspective, using scientific arguments to justify certain conceptions about society, the fight against Darwinism, and notions of harmony and organicity. Written in a flowing literary style, the work was a tribute to erudite learning meant to counter popular misconceptions bred by commonsense beliefs. Observing the social life of animals, Mello Leitão ranked them in a hierarchy where the level of superiority of a species was associated with its ability to engage in social life. Ants belonged on a high plane because they were capable

FIGURE 21. Cover of *A vida maravilhosa dos animais*, by Mello Leitão. Personal files of Regina Horta Duarte.

of friendship, language, tidiness, solidarity, goodness, and, above all, obedience to collective rules. "Following the strictest discipline and the most perfect order," they sacrificed themselves to a brutal regime. Men should learn from this—and since men were superior, their inevitable fate was to make pleasure out of work. Every man was "an anonymous, diligent ant" and should give the best of himself to his home, city, and fatherland and to humanity, until death laid him to rest. Rejecting the idea of conflict, the author underscored the interspecies solidarity demonstrated by birds, elephants, and insects, pointing out that "nature always displays great harmony, where mutual dependence, aid, and, I would go so far as to say friendship, are the general rule."[94] While the Vargas administration promoted a subservient brand of unionism paired with social corporatism, Mello Leitão sang the praises of spiders, which were, "in their patient maneuvers, unflappable worker-bees," "unassuming and quiet." He called attention to the sophistication of monkeys and their societies, grounded in the division of labor, solidarity, feelings of compassion between individuals, and, particularly, "utmost, blind obedience to their leaders." Similar arguments were applied to pelicans, storks, marmots, termites, penguins, sparrows, and many other animals.[95]

In 1940, Mello Leitão published a kind of continuation of this book, *A vida na selva*, which centered on the tropical forest and its plants and animals. One chapter was devoted to the importance of inaugurating parks and nature sanctuaries, another to a compilation of poems and literary excerpts about forests. The book was translated into Spanish and published in Buenos Aires shortly after the author's death, as part of the collection Biblioteca de Autores Brasileños Traducidos al Castellano (Library of Brazilian authors translated into Spanish), which had previously released works by major authors like Pedro Calmon, Oliveira Viana, Euclides da Cunha, and Gilberto Freyre.[96]

As we have seen, Roquette-Pinto, Alberto Sampaio, and Mello Leitão made substantial contributions to the Biblioteca Pedagógica editorial project, and especially the Brasiliana Collection. In tune with the grand nationalist educational endeavor promoted by Fernando Azevedo, their contributions fell within the scope of the National Museum's projects and activities and were commensurate with what they saw as the institution's new role in "national reconstruction." From this perspective, their books should be understood as manifestations of a markedly political praxis.

Their books were also interconnected with their scientific research, as exemplified by Mello Leitão's *Zoogeografia do Brasil*, which clearly lent continuity to Sampaio's *Phytogeographia* and to Roquette-Pinto's theories on Brazilian anthropogeography. They recognized and supported each other as authorities, as we see in Roquette-Pinto's foreword to *A biologia no Brasil* and in Mello Leitão's and Sampaio's painstaking notes and editing of the scientific names of flora and fauna in the fourth edition of *Rondônia*. They defined new fields, worked to launch new disciplines (e.g., Sampaio in his *Biogeographia dynamica*), and laid the groundwork for establishing biology as a stage in advance of natural history, as in Mello Leitão's work. They disseminated the wealth of material stored in the museum's libraries, as in Mello Leitão's three books *Visitantes do Primeiro Império*, *O Brasil visto pelos ingleses*, and *História das expedições científicas no Brasil*. They expanded on debates initiated at major conferences, such as the First Brazilian Eugenics Congress, held in 1929 and discussed by Roquette-Pinto in *Ensaios de antropologia brasiliana*, or the First Brazilian Congress for the Protection of Nature, held in 1934 and analyzed by Sampaio in *Biogeographia dynamica*. Each book "is caught up in a system of references to other books, other texts, other sentences," within a complex web of discourses, where "it is a node within a network."[97]

These books were written in the heat of the scientists' enthusiasm about revitalizing the National Museum and engaging it in school outreach work; they factored into an earnest need to put the institution at the service of popularizing science and to project its scientists as learned authorities well suited to weighing in on the directions the nation should take. What these men put into the Brasiliana Collection also reflected their interest in a collection that was built at the intersection of a number of fields; it was not their goal merely to "invent" readers but also to invent forms of knowledge built at the convergence of disciplines and the crossroads of specialties.

THE RISE AND FALL OF THE NATIONAL MUSEUM IN THE VARGAS ERA

Under the leadership of Edgard Roquette-Pinto, the National Museum earned its place as an institution devoted to producing and disseminating knowledge and as a space for experimenting with new modalities of communication and new scientific and cultural practices. During those years, its members worked assiduously to vanquish any trace of the antiquated image of natural history museums as deposits for exotic items and dusty objects, frequented by eccentric collectors. Looked at from today, many of its initiatives still seem robust and pertinent. The researchers invested in "scientific popularization" that targeted a broad, diversified public, and they wanted ordinary Brazilians to have access to science and art. They offered distance learning, founded educational radio stations, and were pioneers in educational cinema. They were active in the defense of nature, built knowledge at the crossroads of a variety of fields, and fell under the spell of the fascinating potential of communication technologies. They participated in international scientific networks by attending congresses and visiting different institutions while also nurturing correspondence with scholars all over the world. And they endeavored to influence public policy.

For a time, optimism ruled the day. Guided by a "pedagogical illusion" and a firm belief in the redemptive power of scientific reason, men like Roquette-Pinto, Mello Leitão, and Alberto Sampaio were confident their work would help mold a nation befitting their expectations, and they threw themselves into a task that they thought was only "the start of the beginning," in the words of Director Roquette-Pinto.[98]

The museum's scientists were convinced they could find a noteworthy spot for themselves as policymakers in the realms of nature and education, and they labored for the success of their initiatives inside the world of government power. From 1930 to 1934, the Provisional Government threw major support behind these ideas. The scientists were accepted as authorities who deserved to be heard and consulted, as in the case of the bill for the Game and Fish Code or the participation of museum members on major commissions. But they were soon to suffer a series of setbacks.

In July 1934, Roquette-Pinto had to swallow a bitter pill. When the Department of Propaganda and Cultural Promotion opened its doors that year, film censorship passed into the hands of the Ministry of Justice and Internal Affairs, and the *Revista Nacional de Educação* became the ministry's official voice. A few months later, in a letter to the modernist writer Mário de Andrade, Roquette-Pinto classified the act as "one of the worst blows that our government leaders, in their unscrupulousness, have ever to my knowledge struck against the loftiest, purest of ideas." The magazine was the "apple of my eye," "spiritual manna for my poor people." Although the publication was not making money, Roquette-Pinto was thrilled by the letters of praise (over 2,000) that had poured in from all over the country, and the government had received nothing but applause for the initiative. However, he wrote, since the magazine had published "no portrait of living people"—that is, it had not sought favor by highlighting influential personalities—and had not "feted" any of that ilk who "use power to abuse," he saw years of labor, and the fruit of his decades-long dream, suddenly go to waste. The reshuffling in fact signaled the end of the magazine, and not a single issue came out after that. At a time of swift change in the direction of the Vargas administration, the decision had a more profound political significance as well. Communication media were no longer an educational matter but were now subject to new methods of coercion and control by the executive branch, allegedly as a matter of national security.[99] In 1935, Roquette-Pinto stepped down from his post at the museum to join the National Institute of Educational Cinema, where he found an environment conducive to his work on behalf of educational film.

On October 28, 1936, the title of "professor," reserved for department heads, was replaced by the term "naturalist," pursuant to Law 284. In his annual report, Alberto Betim Paes Leme, a geologist and the museum director, lodged a protest against the new law with the education minister, Gustavo Capanema, on behalf of his colleagues at the museum. According to Paes Leme, the exhibits organized by the museum were living lessons for the public at large, as were the classes and lectures offered there. The title of professor placed staff on equal footing with their colleagues at other natural history museums around the world, and now they had been stripped of this "true right."[100] The law was specifically meant to target the museum. Capanema intended to deny museum researchers the status that recognized their efforts to renew natural history and make it an integral part of the biological sciences. The minister used the term "naturalist" pejoratively, belittling the institution as if it were a mere storehouse for collections and its members, antiquarians.

Radio projects were also abandoned. In 1936, having absolutely no funds to sustain it, Roquette-Pinto donated Rádio Sociedade to the government, with the caveat that it remain under the aegis of the MESP. In 1937, he also abandoned the educational radio station PRD-5, at a time when Anísio Teixeira, its cofounder, had become a persona non grata in the political arena, now dominated by Catholic groups and harsh critics of the Escola Nova. Under Roquette-Pinto and Teixeira's original plans, the radio would have extended a vital hand to the University of the Federal District, which was supposed to provide research, teaching, and extension work.

In 1937, Mello Leitão left the National Museum as well and went to teach at the Higher School of Agricultural Science and Veterinary Medicine. His book *A biologia no Brasil*, published that same year, lamented the museum's precarious facilities and collections, the premature demise of the *Revista Nacional de Educação*, the end of many other exciting initiatives led by Roquette-Pinto, and—above all—the way the MESP, with Capanema at its helm, was treating the museum.[101] Sampaio stayed at the museum until retiring in 1941, but he was forced to take a series of leaves for health reasons during his last years, and he wrote much less. Just as dismayed as his colleagues, he bemoaned the uncertain climate that the "naturalists" at the museum faced, including lower salaries and poorer working conditions, all while the professors at the new University of Brazil who had similar scientific and teaching duties were treated to special perks.[102]

Scientists at the National Museum had never constituted a homogeneous group. Roquette-Pinto was of a nonreligious bent, and his views on society and education were rooted in positivist ideals. He fought against both racism and the extinction of indigenous societies and cultures. Although he had been responsible for organizing and coordinating film censorship in the post-1930 era, he battled to keep educational radio and the National Institute of Educational Cinema free from the propaganda intentions of the Vargas administration, coordinated by the Department of Propaganda and Cultural Promotion, which was transformed into the Press and Propaganda Department in 1939. Mello Leitão was a creationist (not unlike most biologists in his day) and a deeply religious anticommunist who had espoused eugenics-based racist positions in the early 1920s. His relationship with Roquette-Pinto, however, softened his stance. He was a colleague of Roquette-Pinto during the founding of the ABE, and although he did not sign the Pioneers Manifesto, he was in practice a loyal defender of the principles of the Escola Nova. Sampaio, for his part, was an admirer of Italian fascism who harbored blatant militaristic propensities, manifested in his obsession with scouting. And none of his writings leave any doubt about his extremely authoritarian tendencies. The men apparently worked around their differences and did not allow them to get in the way of their collective efforts, as attested by the fruitful products of their endeavors. They conducted their activities within the walls of the museum but also worked together with innovative professionals outside the institution, including liberal thinkers like Anísio Teixeira and Fernando Azevedo.[103]

The problems that began to plague the National Museum in 1934, along with the Capanema ministry's sudden denigration of the institution, must be framed within the larger context of the educational question in Brazil and the complex web of political groups that occupied the national stage. When the Provisional Government first came to power, its political platforms and creation of the

MESP had raised great hope for true educational renewal. But as an increasing number of actors entered the fray, other ideals gained strength. Education was an arena of disputes, with the greatest tensions being between Catholic groups and the advocates of secular education. The fight for education was experienced as a political fight for the country's soul. This climate gave rise to very different projects on how to reverse Brazil's backwardness and steer it down the path toward civilization.[104]

Adamantly opposed to the ideals of the Escola Nova, Catholic groups were aggressively making headway in the political realm. While they had initially condemned the movement of 1930, they soon came to see the new conjuncture as an opportunity to undo the radical separation of church and state written into the Constitution of 1891, a legacy of positivist republican activism. Catholics had been reorganizing since 1922, when they had founded the Dom Vital Center, headquartered in Rio de Janeiro. The center's key goals were to bring nonpracticing Catholics into the church and to intercede in debates and initiatives in the national public sphere. Top leaders at the Dom Vital Center, like Jackson de Figueiredo and Alceu Amoroso Lima, argued that the educational reforms underway were prejudicial to the Christian formation of young people. They also criticized the emphasis on scientific training rather than on a humanist education, and accused proponents of the Escola Nova movement of encouraging the advent of a Bolshevik pedagogy, placing boys and girls in the same classroom and promoting equal schooling for students from different social strata.[105]

Rancorous articles published in the magazine *A Ordem*, launched in 1921, pressured authorities to cede Catholics more political influence. They found a strategic ally in the education minister Francisco Campos. He was not a militant Catholic and had in fact been an enthusiastic educational reformer in Minas Gerais, but he felt that support from the church and lay Catholics was strategic to strengthening the new government. In April 1931, an MESP decree allowed public schools to offer religious teaching as an elective, breaking with the secular tradition instituted under the Constitution of 1891. Catholics celebrated the decision while continuing to demand greater space in the Vargas administration. This rapprochement between the government and Catholic interests was not without its paradoxes. Campos's reform of secondary education, initiated in April 1932, displeased Catholic groups, who were critical of the emphasis on technology and science in the teaching curricula, the prevalence of secular premises, and the pedagogical principle of practical education. These groups also attacked the work of Anísio Teixeira and Fernando Azevedo, who held top posts in public education in the states of Rio de Janeiro and São Paulo. They accused these men of being communists and of destroying Christian faith among young people. The same year that Campos enforced his reform, activists from the Catholic movement deserted the ABE, signifying a declaration of war on Escola Nova proponents and evincing the deep divide between the nonreligious and Catholics when it came to education policies.[106]

Brazilian Catholics were jubilant when the name of God was included in the preamble of the new constitution in 1934, and when collaborative ties were re-established between church and state. They also applauded Capanema's appointment as minister of education and public health, which ushered in the development of a new educational project for Brazil and the swift overthrow of Escola Nova principles. Against this backdrop, Anísio Teixeira was openly persecuted. Despite his fondness for U.S. liberal thought and the fact that he had introduced Dewey's educational ideas to Brazil, he was labeled a communist. As director of the Department of Education for the Federal District since 1931, Teixeira had been working to implement far-reaching educational reform from elementary through university levels. As part of this effort, he had spearheaded the founding of the University of the Federal District, which realized the educational dreams of liberal Rio intellectuals. In his July 1935 speech at the inauguration of the university, Teixeira delivered a veritable ode to freethinking as he proclaimed the new institution's commitment to the great liberal and humanist traditions. Alceu Amoroso Lima, powerful head of the Dom Vital Center, took the occasion to write a letter to Minister Capanema in which he declared this to be the straw that would break the camel's back and unleash Catholic discontent.

Those were days of major political upheaval. Vargas was candidly unhappy about the new constitution, which called for elections in 1938 but did not allow him to run. Across the country, workers were going out on marches and strikes. Right- and left-wing radicalization was visible in the ascension of the fascist movement known as integralism and in the communist movement. In March 1935, a small group of intellectuals and military officers formed the National Liberation Alliance (ANL), whose ultimate goal was to fight fascism and imperialism. Luis Carlos Prestes, a member of the Brazilian Communist Party, was nominated president of the alliance, and thousands of people quickly swelled its ranks. The confrontation between the government and the ANL sparked strong anticommunist sentiment. Vargas handed down a National Security Law on

April 4 of the same year, and a communist uprising was crushed on November 23—and used to justify a harsh crackdown by the regime.[107]

The wave of repression reached Teixeira, who was removed from his post at the University of the Federal District. He responded by leaving Rio de Janeiro and taking refuge in his hometown of Caetité, located some four hundred miles from Salvador in an isolated area of the Bahian sertões, where he remained until the end of the Estado Novo in 1945.

Fernando Azevedo also suffered with the rise of the Catholics. In 1935, Amoroso Lima wrote to Capanema to complain about the possibility that Azevedo might be appointed to the General Directorate of Public Instruction. According to Amoroso Lima, Azevedo's technical skills and intelligence were irreproachable, but his presence would serve as a kind of rallying point, because he stood for an educational program that was all too familiar and had already been rejected. In Catholic circles, his appointment would bring nothing but "bewilderment and indignation."[108]

Capanema's reign as minister, from 1934 to 1945, reframed the education question in Brazil. The Catholic Church earned the right to take part in religious teaching at public schools. Gender differences made their way into the educational system: the instruction of girls would focus on domestic skills, while vocational education in business, industry, and agriculture would be introduced for the sons of working-class families. Children of the elite would follow the classic or scientific track upon entering high school, but major emphasis would be placed on the study of languages and on a patriotic, humanist education, to the detriment of the biological, physical, and chemical sciences. Emphasis was put on higher education, which was ascribed the task of training intellectual leaders; in addition, a single teaching standard was to be followed by all universities in Brazil, including the University of São Paulo, founded in 1934 by the state government.[109] Capanema founded the University of Brazil in 1937 and fought against alternative higher education projects. The operations of all universities had to be sanctioned by his ministry. In 1938, he asked Vargas to abolish the University of the Federal District, arguing that it did not meet the benchmark for approval—burying Anísio Teixeira's initiative for good.[110]

It was in this context that the National Museum lost the prestige it had won under the Provisional Government. While its members did not openly challenge Capanema's projects, the museum's renaissance under Roquette-Pinto had been propelled by the ideas of the Escola Nova and belief in scientific teaching, secular education, and dissemination of the knowledge produced at the museum with young patrons, girls and boys alike, through a variety of museum initiatives. The museum had become emblematic of a set of ideals embraced by its "professors," who championed the Escola Nova idea that egalitarian education was a way of offsetting social inequality. Museum scientists had thus taken the opposite tack to Catholic groups, who felt education had "as its purpose, the adaptation of the unequal to a naturally hierarchical social order."[111]

It is worth noting that during the days of hardship, it was a woman who was the central figure of resistance: the anthropologist Heloísa Alberto Torres, museum director as of 1938. While Capanema worked to tailor the education of girls to the development of domestic skills (some high schools offered classes that awarded a "housewife" certificate), Director Torres fought bravely within an institution so bereft of resources that it was sometimes forced to close its exhibit halls to visitors. At the end of the Estado Novo, the museum was attached to the University of Brazil. Currently part of the Federal University of Rio de Janeiro, it still occupies the same building in the Quinta da Boa Vista that harbored the dreams of Roquette-Pinto and his team.

Putting aside their lamentations about the museum's sad situation, Mello Leitão and some of his colleagues launched a promising new initiative: the Society of the Friends of the National Museum. A clear response to the museum's fall from grace, the new society was yet another sign of the vitality and creativity of its members. By then, Mello Leitão had become a successful arachnologist with an enviable scientific résumé and contacts the world over. He still held the respect of the government and in 1937 and 1945 was appointed Brazil's cultural representative on two major international trips. He had illustrious disciples, such as the ecologist and ornithologist Augusto Ruschi, who in 1948, together with the Society of the Friends of the National Museum, founded the Prof. Mello Leitão Museum of Biology, in Santa Teresa, Espírito Santo State, now one of Brazil's centers of biological research. The next chapter will explore the establishment of the society and some aspects of Mello Leitão's later career, particularly his participation in international science circles and his role as an influential educator in the field of biology.

3

THE MAKING OF A BIOLOGIST

Humanity is greatly indebted to zoology, especially from the dawning of this century on, when it shifted from contemplation to activism. Many people still have the idea that the zoologist is a harmless madman, tinkering away with his innocuous, useless obsession. . . . Taxonomy was an indispensable step, but it yielded no immediate fruit. It was only when the field moved to the study of ecology, of the relations of animals with their environments and with other animals, that it became clear what tremendous benefits can be gained from knowing our fauna.

—MELLO LEITÃO, 1943

THE SOCIETY OF THE FRIENDS OF THE NATIONAL MUSEUM

WHEN MELLO LEITÃO INVITED BRAZILIANS TO collect specimens from different regions of the country, pack them properly, and ship them to the National Museum, one of those who answered his call was young Augusto Ruschi, born in the interior of the state of Espírito Santo in 1915, the son of Italian immigrants. As a boy, he spent his days playing with plants and insects on Chácara Anita, the small farm owned by his agronomist father. In 1925, he and his family moved to the city of Vitória, the state capital. His elementary school teacher, the writer and poet Maria Estela de Novaes, had a keen interest in natural history and encouraged her student. Perhaps she was the one who put him in contact with the National Museum, to which Augusto began sending shipments of material he gathered in the woods around the region—material that reached the hands of Mello Leitão. In 1932, the young man sent him boxes of larvae from a pest that had been attacking orange groves. Filippo Silvestri, a zoologist with the agricultural college in Portici, Italy, was one of those who received Ruschi's material through Mello Leitão. In 1937, Silvestri paid a visit to Brazil, and he and Mello Leitão traveled to Espírito Santo to meet the twenty-two-year-old in person.[1]

It was an era of turmoil for the National Museum, as discussed in the last chapter. But some of its members launched a project meant to give the institution a boost of energy. On July 21, 1937, the businessman and philanthropist Guilherme Guinle (1882–1960) met with researchers Mello Leitão, Alberto Sampaio, Paulo Roquette-Pinto (son of Edgard Roquette-Pinto and a naturalist at the museum), Alberto Childe, and Paulo Campos Porto (1889–1968, botanist), among others. Guinle's presence was important and represented much more than the financial support of his millionaire family. He was a pioneering entrepreneur, an "enlightened" nationalist, and a sympathizer of the National Liberation Alliance, whose participation had been decisive during early twentieth-century modernization projects in Rio de Janeiro; he had also had a steady hand in the definition of public health policies for the poor. His brother Carlos Guinle (1883–1969) had worked alongside Roquette-Pinto at Rádio Sociedade as a board member in the 1920s.[2]

The meeting attendees decided to establish a society to help support the museum—along the lines of the Société des Amis du Muséum National d'Histoire Naturelle in Paris—and start a magazine. They also discussed the bylaws of the new organization, drafted by Mello Leitão. Approved and published that same year, they stipulated that the society would safeguard the cultural heritage of the National Museum in five basic ways: by helping to enrich its collections and library; by keeping private collections from being passed to foreign institutions; by offering assistance to expeditions of Brazilian and foreign naturalists, as long as they ultimately added to the museum's collections; by promoting the establishment of zoos and reserves for flora and fauna; and by doing its utmost to foster better knowledge of Brazilian flora and fauna and respect for the nation's indigenous peoples with a view to preventing their extinction. The association was open to anyone interested, and it defined very flexible levels of membership fees, ranging from five mil réis a month up to twenty contos de réis or more, payable in cash or through donations of collections or books.[3]

Guilherme Guinle was appointed president; Mello Leitão, vice-president; and Campos Porto, secretary. The first and only issue of the magazine *Uiára* came out in late 1937. Its title harkened back to Roquette-Pinto's reference to the legend of Uiára in his book of essays *Seixos rolados*. In that essay, Roquette-Pinto had said that Brazil held a secret of nature: the song of Uiára, seductive and irresistible, sensed in the country's "mountains and valleys, in its forests and rivers," where "beautiful things" stir great and deep love. Whoever caught a glimpse of Uiára—the idealization of Brazil—would not be able to resist her and would be cast into the depths of "her charms" and lost in the "love of her wonders."[4] In his introduction to the first issue of the society's magazine, Roquette-Pinto declared his enthusiasm for the National Museum, "a miniature of the Fatherland," to which he had devoted decades of his life. He praised Guinle, "an illustrious brasiliano who will leave behind a legacy of patriotism bound up with lofty initiatives and achievements."[5] Paulo Roquette-Pinto then offered a brief historical overview of the institution, highlighting the period when his father had been director, from 1926 to 1935, and making special mention of the Assistance Service for the Teaching of Natural History and its activities. The issue also contained nine popular science articles, signed by Mello Leitão, Alberto Sampaio, and Alberto Childe, among other National Museum scientists.

The magazine was printed on high-quality paper and featured black-and-white photographs with excellent resolution, as well as a number of drawings. Two images appear on the cover: on the top half, a mythological creature, part animal, part human, wades through a lake surrounded by dense vegetation—a brasiliano landscape. The title appears to be submerging into the water, like the readers who are expected to dive in, fascinated by nature. A 3-D cross-section drawing of an agate rock is on the bottom half, its deep-layered terraces reminiscent of a miniature cave (figure 22).

In regard to the society's goal of creating nature reserves, Mello Leitão had one very specific proposal. As mentioned earlier, the visit of the Italian entomologist Filippo Silvestri had prompted Mello Leitão to go to Santa Teresa to meet Ruschi, his young correspondent, who came up with the idea of setting up a biological station in the region. In 1939, Ruschi donated fifteen contos de réis to the Society of the Friends of the National Museum to buy a 346-acre lot in an area where he was doing important research on orchids; the flowers were in urgent need of protection, and this justified the establishment of a small biological station. Mello Leitão celebrated the initiative in the pages of his book *A vida na selva*, where he sang the praises of the society for purchasing "a small but highly interesting fauna and flora reserve." The acquisition required lengthy negotiations and involvement by the museum's director, Heloísa Alberto Torres.[6] Ten years later—and a year after Mello Leitão's death—Ruschi founded the Prof. Mello Leitão Museum of Biology on the property and began publishing the journal *Boletim do Museu de Biologia Mello Leitão.*

The Society of the Friends of the National Museum disbanded in the early 1940s.[7] The building that housed the museum was deteriorating badly and had

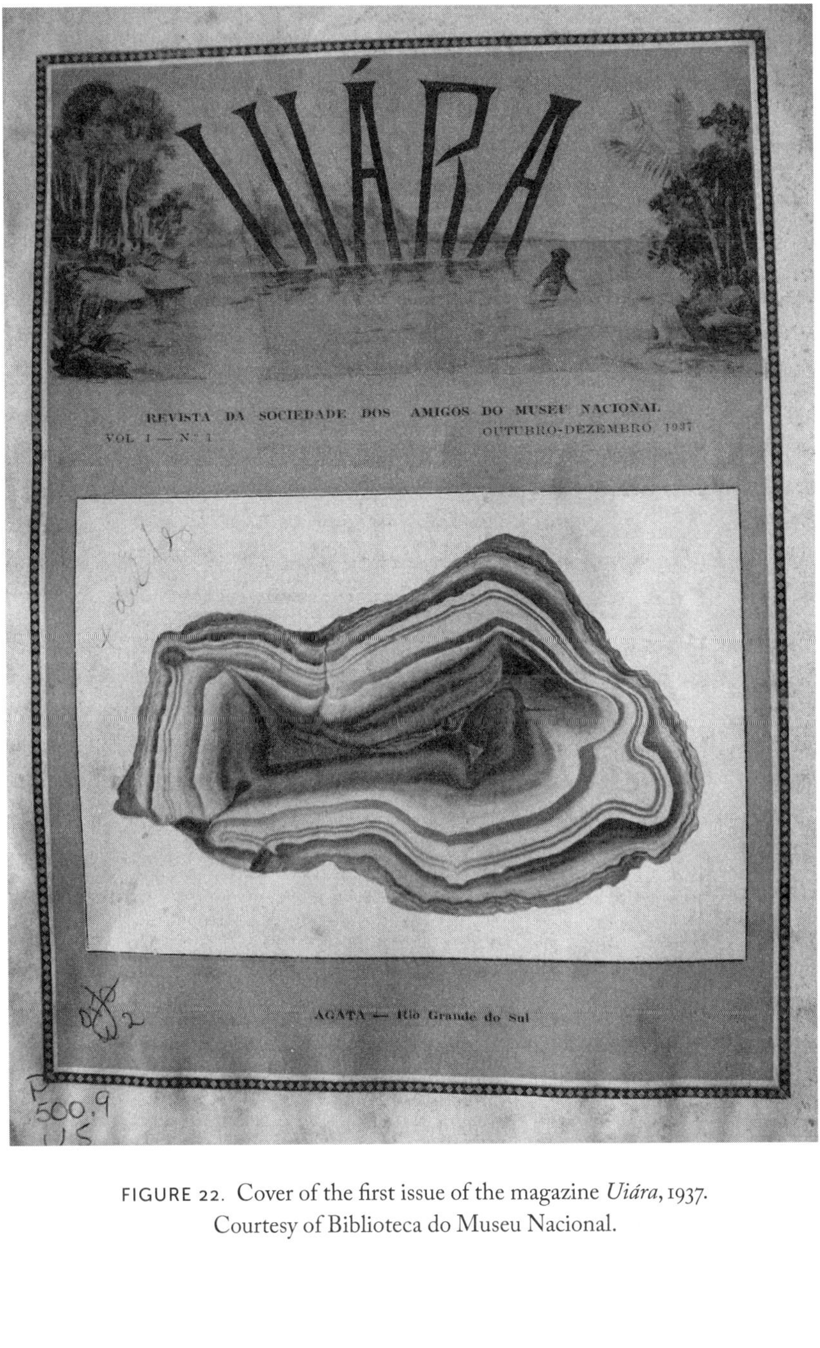

FIGURE 22. Cover of the first issue of the magazine *Uiára*, 1937. Courtesy of Biblioteca do Museu Nacional.

to be closed for repair in 1941. In 1944, a fire caused extensive damage. In 1946, a short piece in the newspaper *A Noite* decried the situation: in the past, the museum had been "an important research center, served by figures as learned as a Roquette-Pinto, a Mello Leitão, a Professor A. J. Sampaio." The paper also mentioned the museum's noteworthy publications and the many visits paid by students during its golden age. It still held a vast collection of documents and material invaluable to science teaching. There one could find "a summary of our fauna, our flora, our mineral resources, and documentation on our aborigines"—in short, the museum was a priceless treasure and its state of utter neglect, a crime. The newspaper said the museum had morphed into "a ghost house, its doors closed for over a year and a half," the victim of the lack of interest of the Ministry of Education and Public Health (MESP) under Capanema. An appeal was made to the acting minister, physician Ernesto de Souza Campos, asking him to attend to this "plaintive case, above all for those who had known and admired the excellence of this organization in the past."[8]

The three personages chosen by *A Noite* to evoke the glory days of the National Museum—whose careers had intersected in the 1920s and 1930s, as we saw in the first chapter—had by then become distanced from the museum. Sampaio would die before the end of 1946, following a long illness. Shortly after helping found the Society of the Friends of the National Museum, together with Mello Leitão, he took a leave of absence that stretched into all of 1938, and in 1941 he retired. He did, however, still find time to publish some major works. In 1942 he translated into Portuguese three chapters on botany from Georg Marcgrave's *Historia Naturalis Brasiliae*, with commentaries of his own, and in 1944 he released *A alimentação sertaneja e o interior da Amazônia*, as mentioned earlier.[9] Mello Leitão paid tribute to his "dearly departed friend" in a lengthy speech before the Brazilian Academy of Sciences (ABC) given on the first anniversary of his passing. He reminisced about their joint initiatives, dating to the 1920s, and talked a bit about the botanist's career abroad.[10]

Roquette-Pinto had transferred to the recently established National Institute of Educational Cinema (INCE), where he was a leading figure from 1936 to 1946. He remained heavily committed to projects to transform Brazil through the use of media and invested much energy in educational film and the blending of culture, art, and technology. Once installed at the INCE, he resumed many of the projects initiated with the publication of the defunct *Revista Nacional de Educação*. Several factors converged in his favor. For one thing, in 1934,

Vargas had proclaimed cinema a "book of shining images" through which people would learn to love Brazil and have faith in the destiny of their fatherland. For another, Capanema's establishment of the INCE was part of a strategy to strengthen his ministry, a "way to cement its role within the government."[11] The minister thus pulled educational cinema back into his decision-making orbit. He appointed Roquette-Pinto to head up the new agency because he had been a pioneer in the field since his time at the National Museum in the late 1920s.

In 1936, Roquette-Pinto traveled to Europe, at Capanema's behest, to observe educational cinema in countries like Germany, France, and Italy—the latter was a forerunner in the area, and Roquette-Pinto had been corresponding with Luciano de Feo since 1927.[12] A total of 252 educational films were made during Roquette-Pinto's tenure at the INCE, and he had a direct hand in many of them, including three in 1937 (about the rabies vaccine, the skies over Brazil, and the *Victoria amazonica* plant), two in 1939 (about the yellow fever vaccine and the electric eel), and one in 1942 (about physical and chemical experiments). He also worked with filmmaker Humberto Mauro on the production of three other movies, whose themes were the discovery of Brazil, the early adventurers of Brazil known as *bandeirantes*, and the ceramics of the Marajó Island indigenes.[13]

In the year that Mello Leitão helped establish the Society of the Friends of the National Museum and was elected its vice president, he was enjoying a growing reputation as a zoologist with expertise in arachnids, both at home and abroad. His participation at a recent congress in Mendoza, Argentina, had earned him notable attention. He was also part of an intricate web of researchers and institutions that were accumulating knowledge of arachnids and furiously publishing scientific papers and articles for the public at large. In 1937 alone, he released three titles in the Brasiliana Collection, fourteen papers in scientific journals, and two articles in popular magazines. This was the same year he quit his job at the National Museum. Following the November 1937 coup, Decree-Law 24, of November 29, prohibited individuals from holding more than one paid public position, and Mello Leitão opted to stay at the Higher School of Agricultural Science and Veterinary Medicine, probably because he thought it would be the most advantageous place for his scientific work. In 1938, he was named an honorary member of the National Museum, and for the rest of his life he kept in touch with the institution through his longtime disciples in the study of arachnids. He also remained an active member of the Society of the Friends.

In later years, he taught zoology at the University of Brazil in the philosophy (1939–41), veterinary medicine (1941–46), and agronomy (1946–48) programs. He also served a term as president of the ABC from 1943 to 1945.

Of the three great enthusiasts during the golden age of the National Museum, Mello Leitão was the only one to forge a true career as a scientist and find acceptance in Brazilian and international academic circles. Convinced of the potential of biogeography—a topic to which he was particularly devoted in the final years of his life—Mello Leitão embraced a Pan-American ideal and won esteem beyond the borders of Brazil. In an era when there were no courses specific to the field in Brazil, Mello Leitão made his name as a biologist.

In my estimation, by the first half of the twentieth century we are justified in speaking of biology as a field of scientific knowledge in Brazil. I have defended the hypothesis that biology was an influential player in Brazil's political life at a time when the term "natural history" still ruled the study of fauna and flora.[14] Biology came into being as a field at a time when scholars dedicated to nature and especially to the study of living organisms were conducting research and disseminating their findings, while at the same time endeavoring to respond to the challenges of the day and always in dialogue with a series of historical and social transformations. According to these scientists, dried specimens, insects, and taxidermied animals arranged in cabinets and on shelves at the National Museum were just the first stage of knowledge, to be followed by a more complex approach, framed as the study of life: *bio + logos*. They felt that natural history, with its predilection for collecting specimens, was incapable of answering the questions then on the agenda and represented merely a preliminary step in the tasks they had set and the role they saw themselves fulfilling in society.

This chapter will explore the international networks in which Mello Leitão's participation as an arachnologist intersected with Brazil's policies in foreign affairs. This is followed by a discussion of the pinnacle of his career in Brazil—his election as president of the ABC—and his continued work in science communication through the writing of textbooks. As a researcher, Mello Leitão combined the best of academic scholarship with a deep concern for the education of high school students and future teachers at normal schools. Always, Mello Leitão approached the practice of biology as an activist, believing not just in the transformative nature of knowledge but also in the strategic role he believed should fall to biologists. He reaffirmed the notion of a republic of scientists who were reliable guides for government leaders and for the governed, at a time when men like him were envisioning an age of biology.

I have chosen to focus on Mello Leitão for specific reasons. His career stands as a thought-provoking reflection of the complex situation of natural history within the broader field of biology during the first half of the twentieth century. Interestingly enough, although he criticized mere collecting of specimens and pure and simple classification, he himself was always a tireless collector and zealous taxonomist. Mello Leitão acquired great renown in zoology without ever discarding natural history. His approach to the field was dynamic, with a view to its renewal; he kept well abreast of the changes in his area of expertise, especially those regarding zoogeography and ecology.

In 1937, when the National Museum began grappling with major challenges and its staff members were deprived of their scientific status, Mello Leitão opted to resign from his post, despite his declared unconditional esteem for the institution. It may well be that he believed that in this way he could achieve peer recognition and avoid being a target of the disdain that so many then felt for natural history and those devoted to it. In order to understand how Mello Leitão made a name for himself as a biologist among his contemporaries, we must look closely at the variety of strategies he employed, his scientific relationships, and his activities; we must also follow his footsteps well beyond the walls of the National Museum.

BIOLOGY WITHOUT BORDERS

In 1949, the Argentinean collection Biblioteca de Autores Brasileños released its twelfth volume, *La vida en la selva*, a translation of Mello Leitão's 1940 title for the Biblioteca Pedagógica. The introduction was written by Federico Daus (1901–88), a leading Argentinean geographer and professor and the author of a number of classic books about his country. The publication was posthumous, since Mello Leitão had passed away in December 1948 following a serious illness (figure 23). Daus lamented the fact that one of "the most prominent names in the intellectual world," a scientist of "remarkable universal renown" respected by the greatest Argentinean entomologists, and the author of a "bountiful and polymorphous" set of works had vanished from the scene when his book was nearly ready. After presenting some biographical information about Mello Leitão, Daus went on to extol both the scientific value of *La vida en la selva* and its literary merit, poignant tone, and wise, sensitive descriptions.[15] Until then, Mello Leitão was probably unknown to Argentina's general reading public, but his name was

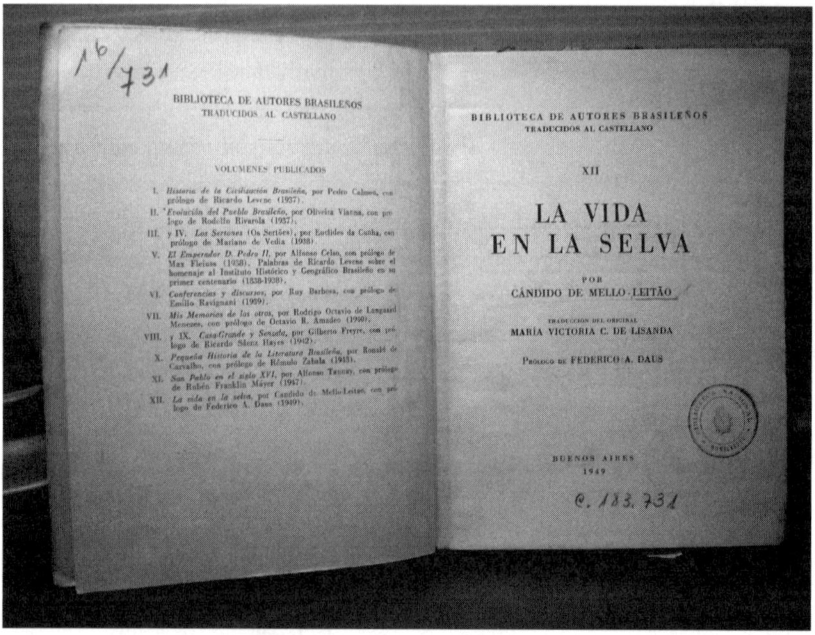

BIBLIOTECA DE AUTORES BRASILEÑOS
TRADUCIDOS AL CASTELLANO

VOLUMENES PUBLICADOS

I. *Historia de la Civilización Brasileña*, por Pedro Calmon, con prólogo de Ricardo Levene (1937).
II. *Evolución del Pueblo Brasileño*, por Oliveira Vianna, con prólogo de Rodolfo Rivarola (1937).
III. y IV. *Los Sertones* (Os Sertões), por Euclides da Cunha, con prólogo de Mariano de Vedia (1938).
V. *El Emperador D. Pedro II*, por Alfonso Celso, con prólogo de Max Fleiuss (1938). Palabras de Ricardo Levene sobre el homenaje al Instituto Histórico y Geográfico Brasileño en su primer centenario (1838-1938).
VI. *Conferencias y discursos*, por Ruy Barbosa, con prólogo de Emilio Ravignani (1939).
VII. *Mis Memorias de los otros*, por Rodrigo Octavio de Langaard Menezes, con prólogo de Octavio R. Amadeo (1940).
VIII. y IX. *Casa-Grande y Senzala*, por Gilberto Freyre, con prólogo de Ricardo Sáenz Hayes (1942).
X. *Pequeña Historia de la Literatura Brasileña*, por Ronald de Carvalho, con prólogo de Rómulo Zabala (1943).
XI. *San Pablo en el siglo XVI*, por Alfonso Taunay, con prólogo de Rubén Franklin Máyer (1947).
XII. *La vida en la selva*, por Candido de Mello-Leitao, con prólogo de Federico A. Daus (1949).

BIBLIOTECA DE AUTORES BRASILEÑOS
TRADUCIDOS AL CASTELLANO

XII

LA VIDA
EN LA SELVA

POR
CÁNDIDO DE MELLO-LEITÃO

TRADUCCIÓN DEL ORIGINAL
MARÍA VICTORIA C. DE LISANDA

PRÓLOGO DE FEDERICO A. DAUS

BUENOS AIRES
1949

FIGURE 23. Argentinean edition of *La vida en la selva*, by Mello Leitão. Courtesy of Biblioteca Nacional de Uruguay.

certainly not unfamiliar in academic circles, since he had long engaged in productive exchange with various institutions and researchers there, as well as in other Latin American and European countries and in the United States.

In "O livro de minha vida" (The book of my life), Mello Leitão's handwritten autobiography—which he intended to be read by his grandchildren—he offered a retrospective evaluation of his career. As he wrote there, his peers held him in the highest regard, and he had been paid tribute many times over by scientists from Argentina, England, Germany, Brazil, Switzerland, Uruguay, Italy, and Latvia who had named nine new genera and twenty-three new species in his honor (figure 24a and b). He was a prodigious writer, having published fifteen papers in the field of medicine, 117 popular science articles in magazines and newspapers, 212 specialized zoology texts for scientific journals, ten biology textbooks, three theses, four book translations, a biology glossary, and, in the Biblioteca Pedagógica, five books for the series Brasiliana and two for the series Scientific Training. He had debuted in arachnology in 1915 with an article in

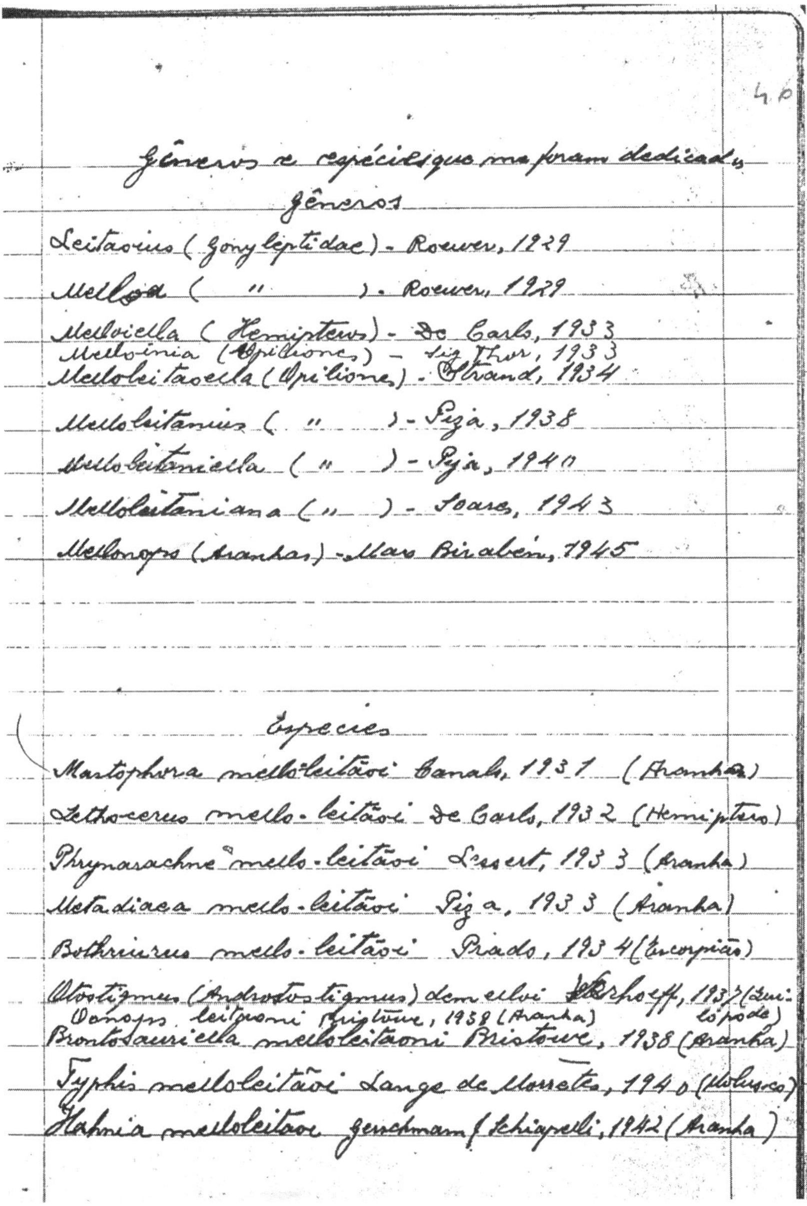

A

FIGURE 24A AND B. List of genera and species named in honor of Mello Leitão by other scientists. Handwritten by Mello Leitão in "O livro de minha vida." Courtesy of Academia Brasileira de Ciências.

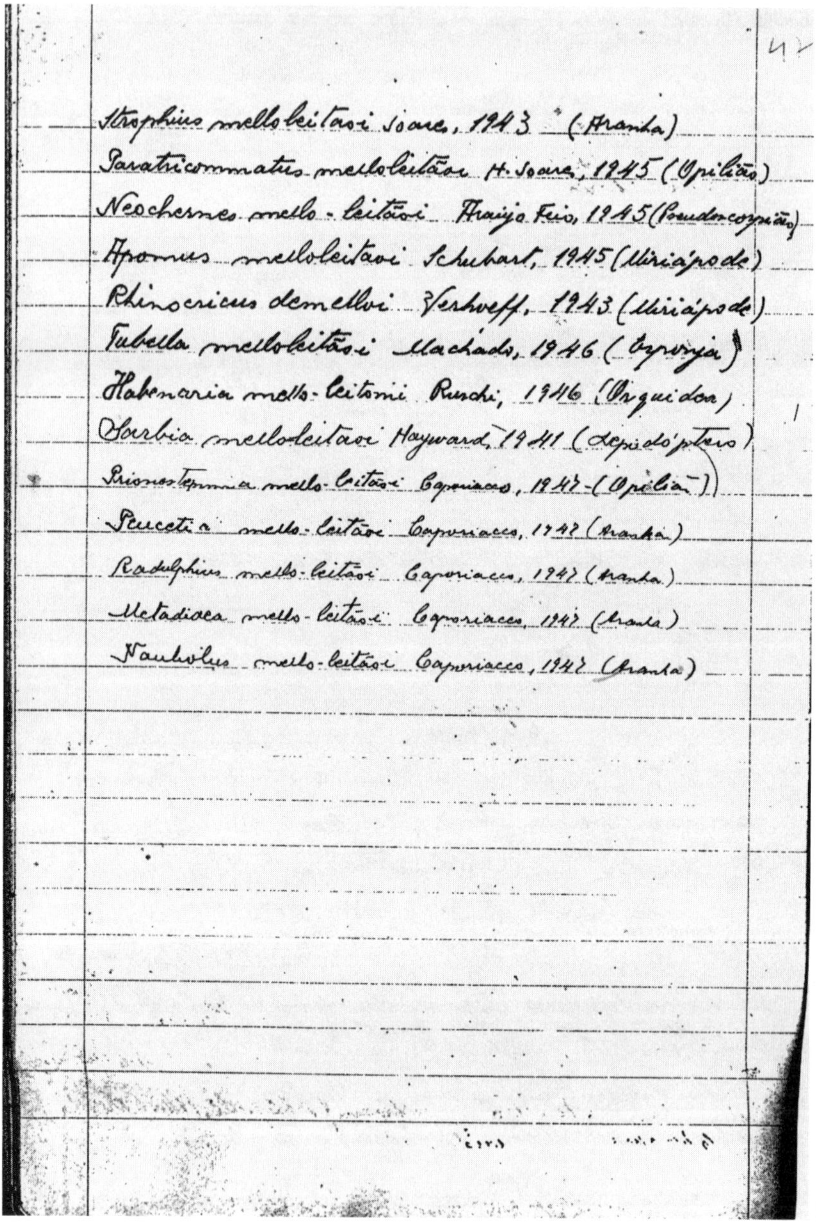

B

FIGURE 24A AND B. (*Continued*)

the Zoology series in the magazine *Brotéria* (published in Braga, Portugal), in which he took the opportunity to urge readers to mail him their own arachnid specimens.[16] Over the years, he kept in touch with other researchers, exchanged specimens, and was a corresponding member of countless scientific institutions. Mello Leitão built up an admirable network of relationships, through which he earned admiration in various corners of the world. His career is exemplary not only because of its trajectory in specialization but also as a reflection of the construction of biology as a specific field in Brazil. He began his work in medicine and then moved to the crossroads of natural history, education, and biology, eventually cementing his standing as a biologist specializing in arachnids and as a university professor of zoology.

His attendance at conferences and his participation in exchange programs figured large in this process. As we saw in the first chapter, Mello Leitão went to Europe in 1926 to intern at pediatric clinics and at the National Museum of Natural History in Paris, where he met top zoologists, many of whom he corresponded with later.[17] As important as this trip was to Mello Leitão's career, it was his recruitment to the National Museum in Rio de Janeiro that would prove truly decisive in his relations with foreign researchers and institutes. Shortly after he joined the museum (exempted from taking the qualifying exam because he had fulfilled an alternative prerequisite, the publication of outstanding scientific works),[18] Mello Leitão was appointed to represent his country during the first round of a cultural exchange program between Brazil and Uruguay.

The program grew out of a treaty signed in 1918 to recover a debt that Uruguay owed Brazil. A bridge was to be built between the two countries (work began in 1927, and the bridge was inaugurated in 1930), while other monies would be allocated for trips aimed at tightening cultural ties.[19] The first Brazilian commission, appointed by the MESP, comprised Humberto de Campos (member of the Brazilian Academy of Letters), Rosalina Coelho Lisboa (journalist and writer), Armanda Álvaro Alberto (member of the ABE),[20] Renato Pacheco (physician and president of the Brazilian Sports Confederation), Ernani Lopes (president of the Mental Hygiene League), and Mello Leitão, then of the National Museum and the Rio de Janeiro Normal School.[21]

Mello Leitão arrived in Montevideo on October 23, 1931, charged with the chief mission of observing high school education in Uruguay. He visited many preparatory, secondary, and normal schools, plus teachers associations. He also taught classes at the University of Montevideo, gave a lecture on the "wonderful life of spiders," and was compared to Fabre and Maeterlinck by the Uruguayan press.[22] During all his visits, Mello Leitão drove home the need for stronger intellectual relations between Brazil and Uruguay as well as increased interaction in the realms of science, literature, politics, and education. He referenced Brazil's early twentieth-century foreign policy under the diplomat Rio Branco, when sights had been set on continental solidarity.[23] Mello Leitão said he was delighted by everything he saw and that he had arrived in Uruguay as a Brazilian but would return to Brazil wholly Uruguayan. He extended this desire for closer relations to other countries in South America too, countries that often knew nothing about each other because they were more attached to Europe than to their neighbors. An effort had to be made to work toward the "invaluable goal of forming great intellectual families" and toward "an exceptional future for the new America," where "brothers" would match the feats attained in other continents.[24]

Mello Leitão backed the Provisional Government's general foreign policy, which at first largely continued the Pan-American policies that Rio Branco had put in place in hopes of overcoming Brazil's isolation within Latin America, especially with regards to the Plata River region. The same newspapers that followed the entourage of Brazilian scholars in Montevideo, Buenos Aires, and La Plata reported on the conference between Argentina, Brazil, and Uruguay that was held in Montevideo a few weeks later, in December 1931, to promote favorable foreign trade conditions for livestock exports from the three countries.[25]

The trip meshed well with some of Mello Leitão's longstanding pursuits. Since 1920, he had published eight papers on South American arachnids, in addition to his numerous studies on Brazilian species. He had established contact with the Bernardino Rivadávia Museum of Natural History in Buenos Aires and had received several Argentinean specimens for classification. In 1929, at an event held at Brazil's National Academy of Medicine, he met entomologist Ergasto Cordero, from Montevideo, and physiologist Bernardo Houssay, from Buenos Aires. Thanks to his 1931 exchange trip, he had the opportunity to deepen these bonds, and it was with this in mind that he headed to Buenos Aires at his own expense as soon as he was done in Uruguay. In the Argentinean capital, the Bernardino Rivadávia Museum of Natural History honored him at a special session, where he gave a lecture on the zoogeography of South American *Opiliones*, an arachnid order commonly known as daddy longlegs, or harvestmen. He met a number of entomologists there, most notably José Canals, and initiated what were to become rewarding collaborations with him and other researchers.[26]

In 1935, Mello Leitão traveled to Europe, having been appointed by the Ministry of Agriculture as Brazilian envoy to the Sixth International Congress of Entomology in Madrid and the Twelfth International Congress of Zoology in Lisbon. The Ministry of Agriculture no doubt sponsored his participation because the government had come to see the value of the practical applications of entomology to agricultural problems, as discussed in chapter 1. Mello Leitão's proposal to hold the Thirteenth International Congress of Zoology in Rio de Janeiro was approved at the event in Lisbon. Unfortunately, nothing ever came of it, because, as he said, there was a lack of interest back in Brazil, and so the country missed out on a "magnificent opportunity."[27]

Not long after this trip, Mello Leitão received several honors, attesting to his rising acclaim. He was made a corresponding member of the Chilean Academy of Natural Sciences (1936), the La Plata Museum of Natural Sciences (1936), and the Societé Scientifique du Chili (1937). The Natural History Museum of Basel sent him collections of spiders for classification, while the Natural History Museum of Barcelona shipped him some daddy longlegs.[28] The trip had yielded another bonus: the arachnologist was able to attend a series of discussions on zoogeography, further piquing his interest in the topic. In 1937, he published *Zoogeografia do Brasil* as part of the Brasiliana Collection; a revised, expanded version was released in 1947, a product of the author's further studies in the field.

In April 1937, Mello Leitão was assigned by both the Ministry of Agriculture and the National Museum to attend the Second Meeting of the Natural Sciences, in Mendoza, Argentina. His participation secured him a definitive place in the international arachnology network, in addition to strengthening his connections with the Latin American scientific community.

The largest number of papers at the conference were presented on the zoology of invertebrates. While Mello Leitão was not the only Brazilian present, he organized his country's contributions and was truly in his glory at the meeting. He was the first to speak and gave a paper on the zoogeography of Argentinean spiders. He read another four papers at other gatherings and, according to the minutes, was the sole attendee whose presentations on the zoology of invertebrates received "lengthy applause."[29]

At the close of the meeting, Mello Leitão was named to represent the La Plata Museum of Natural Sciences at the Seventh International Congress of Entomology, scheduled to be held in Berlin in 1938. Tensions with Germany notwithstanding—1938 would prove a horrific year as the Nazis intensified their persecution of the Jews and their expansionist efforts—this appointment was a special tribute and won Mello Leitão recognition as one of the leading experts on South American spider species. His scientific collaborations accounted for the description of 450 new species in Argentina alone; when he passed away in December 1948, this figure represented half of the known spiders in that country.[30]

The decision to name Mello Leitão to represent the La Plata museum at the event in Berlin came at a moment when Brazilian foreign policy was leaning toward rapprochement with Germany. The minutes of the meeting in Mendoza fail to provide any insight into what Mello Leitão himself or any other researcher in attendance thought about his selection, but the choice must have been very well received by the Brazilian government. Trade relations between Brazil and Germany had been growing since 1934. In 1935, Germany and Brazil had begun working together to fight communism. The Vargas administration wavered between aligning with the United States—which stood for free trade practices and the pursuit of liberal democracy—and Germany. "Poor in foreign currency," Germany "wanted to gain ground in the Latin American region" by bartering goods and "promoting nationalist authoritarianism."[31]

Relations between Brazil and Germany were strong in April 1937 when Mello Leitão was named to this mission. A few months later, the German press celebrated Brazil's November 1937 coup, while relations between Brazil and the United States suffered. But Brazil's flirtation with Nazi Germany soured in the following months. Oswaldo Aranha, named minister of foreign affairs in March 1938, was resolute about firming up relations with the United States and demanded that Vargas steer Brazil in this direction. He found a willing ear in the Brazilian dictator, annoyed as he was about the news of Nazi demonstrations in southern Brazil and about Germany's plans for the region, home to many German immigrants. In 1938, there were increasing signs that the German government would enforce a policy of protecting its citizens wherever they might be, and this displeased the Brazilian government. The Nazis had organized a Brazilian branch, with its main offices in São Paulo and representatives in other states with sizable clusters of German immigrants. In April 1938, Vargas began enforcing a series of measures in response: foreigners were banned from political activities in Brazil, education was nationalized, Portuguese became the official language in Brazilian schools, and teaching establishments were forbidden to accept foreign financial aid.[32] This was the backdrop against which Mello Leitão's trip to Berlin did *not* take place. He nevertheless

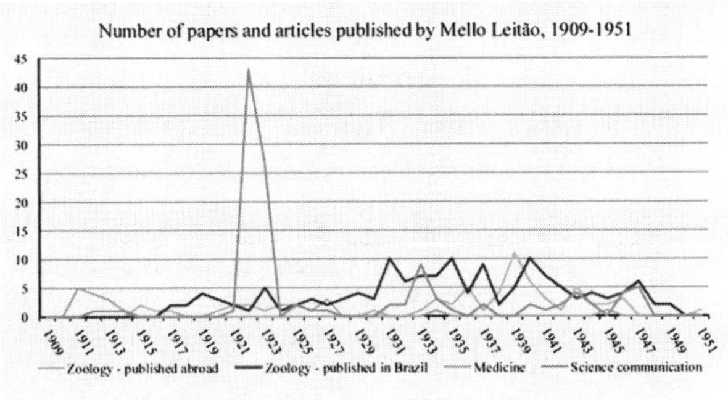

FIGURE 25. Number of papers and articles published by Mello Leitão, 1909–51. Graph by Regina Horta Duarte. Data drawn from Mello Leitão, "O livro de minha vida"; and Kury and Baptista, "Arachnological Papers."

published a paper in the annals of the event, and his appointment is mentioned in his biographical data on file with the ABC.[33]

In Mendoza, Mello Leitão had adopted a rather ambivalent stance—not unlike Brazilian foreign policy during those years. He accepted the assignment in Berlin but in interviews and at social events, he consistently emphasized the need to forge ties within Latin America. He declared that he was enthralled with everything: he extolled Argentina's nature, the beauty of the city of Mendoza, the fine organization of the event, the merit of the researchers in attendance, the fertile land and its plentiful viniculture, and the importance of the wine industry. He ended his talk by expressing an "ardent wish for American fraternity."[34]

The trip yielded professional returns. Not long after, Mello Leitão would be welcomed as a corresponding member of the Valparaiso Scientific Society (1938), the Colombian Academy of Exact, Physical, and Natural Sciences (1939), and the National Academy of Sciences in Córdoba, Argentina (1940). During this period, his scientific publications also reflected a growing reputation and intense devotion to his work. As shown in figure 25, the zenith of his publications in zoology came in the 1940s, when his papers appeared regularly in Brazilian journals. His zoology papers in foreign journals reached their highest number shortly after his attendance at international conferences in 1935 and 1937, testimony to how these events figured into his growing prominence. His publication of popular science articles hit a peak during the days of the *Revista Nacional de Educação*, although still falling short of the levels reached in 1922–23, when he wrote a science column for the Rio newspaper *O Imparcial.* Mello Leitão left the National Museum and opted for the Higher School of Agricultural Science and Veterinary Medicine the same year that he returned from Mendoza. His absence was no doubt one of the factors behind the museum's decline. Edgard Roquette-Pinto had already left, and now the institution was losing one of its most famous researchers at home or abroad and one who had been a tireless collaborator with the Assistance Service for the Teaching of Natural History.

Mello Leitão was steadfast in his relationship with the museum throughout his life. He consulted the library often, published in the museum's journal, collaborated regularly with his former disciple, José Lacerda de Araújo Feio (likewise an entomologist, Feio became head of the museum much later, in 1967), and always made sure to send several offprints of his published papers to the institution. In 1938, the museum served as intermediary between Mello Leitão and Britain's Natural History Museum, which shipped the expert a collection of *Proscopiidae* insects for classification. In 1944, during one of the darkest periods in the history of the museum, he sent it a Christmas gift: some specimens of arachnid species not yet among its collections. The accompanying card, addressed to Director Heloísa Alberto Torres, was written in an affectionate tone and signed "old friend."[35]

The researcher also left some of his personal belongings at the museum. Feio's archive holds two bound albums, one containing newspaper clippings of articles written by Mello Leitão from 1922 to 1926; they have been carefully pasted in and dated, though most do not indicate place of publication. In the second album, the compiler cut and pasted newspaper clippings of interviews or reports about the arachnologist; these are in chronological order, with date and place of publication annotated. The clippings range from 1925 to 1945, which tells us that either Mello Leitão himself or a member of his family left the book at the museum long after his departure, with the conscious intent of documenting his career and, more importantly, bearing witness to the weight of his work.

The second album holds many news reports and photographs from his final trip abroad, in 1945. Berlin had been seized from the Nazis, and Japan had just announced its surrender when the Brazilian committee, Mello Leitão included, arrived in Montevideo on September 3. Like his previous visit, in 1931, this trip was part of the Brazil-Uruguay cultural exchange effort inaugurated by the 1918 bilateral treaty. Mello Leitão had participated in the first mission, and now he joined this one—the last of the Vargas era, since the ruler would be deposed a few weeks later, on October 29. Oswaldo Aranha had resigned as minister the previous year in protest over the Brazilian government's repressive policies, and at the time of Mello Leitão's trip, the Estado Novo dictatorship was rapidly losing face.

There were three in the group this time. Accompanying the arachnologist were the poet Maria Eugênia Celso, granddaughter to famed Brazilian poet Afonso Celso, along with Elmano Cardim, director of the *Jornal do Commércio*. The newspapers in Montevideo heralded this as the most brilliant mission to date. For the first time in the history of the republic, the Uruguayan Senate received intellectuals who did not belong to its body of parliamentarians, and Mello Leitão's address was included in its annals. The three Brazilians were also honored at a special session of the private cultural institute known as Atheneo, which had no connections with either state or church and was a pioneer in the innovation of secular education in Uruguay.[36]

Mello Leitão gave lectures on scientific exploration in the Brazilian Northeast at the Uruguayan Historical and Geographical Institute, on new directions in biogeography at its National Museum of Natural History, and on araneism (arachnidism) at the Greater University of the Republic School of Medicine. He was received by a number of authorities, including the greatest Uruguayan biologist of the day, Clemente Estable.[37] The Museum of Natural History and the Historical and Geographical Institute paid him special tribute. Brazil's Ministry of Foreign Affairs had tasked Mello Leitão with evaluating the promotion of Brazilian scholarship inside the borders of its southeastern neighbor. And, once again, the researcher extended his trip to include Buenos Aires, where he was likewise received with honors and his portrait was unveiled at the Bernardino Rivadávia Museum of Natural History. He lectured at the La Plata Museum of Natural Sciences, the University of Buenos Aires School of Exact, Physical, and Natural Sciences, and the Argentina Entomological Society, and he met with other zoology professors from Argentina.[38]

At the time, Brazil's relations with Latin American countries were robust, thanks to the tenacious Pan-American policy enforced by the Vargas administration, supplanting the uncertainty characteristic of the early years of the Estado Novo and its strategy of "pragmatic equidistance."[39] In January 1942, Brazil abandoned its neutrality and officially broke off relations with the Axis, aligning instead with the Allies. In August 1942, it declared war on the Axis and in 1944 sent troops of the Brazilian Expeditionary Force to fight in Italy.

In the cultural realm, Brazil had already initiated a sure and steady push to draw closer to other countries of the Americas, as reflected in the 1941 launching of *Pensamento da América* (American thought), a supplement to the Estado Novo's official newspaper, *A Manhã*, both of which were published as part of "a joint project by the Press and Propaganda Department and intellectuals from the modernist movement who held key posts in Vargas politics."[40] The supplement was meant to foster cultural relations among South American countries. It featured articles on an array of topics, like literature, music, history, politics, and folklore, all intended to redraw the image of America and bring Brazilians closer to their neighbors. The supplement was a component in a broader movement for international cooperation and exchange, which stimulated "trips, book translations, scholarships, language courses, art exhibits, and concerts." It also worked to establish chairs at universities and special sections within libraries and to encourage exchanges of books.[41]

The desire to strengthen bonds, as exemplified by *Pensamento da América*, was not without precedent in Latin American foreign policy. In 1936, the Inter-American Conference for the Consolidation of Peace, held in Buenos Aires, had called for the exchange of educational films, publications, and works of art. The Biblioteca de Autores Brasileños was launched in 1937 (as mentioned earlier, its twelfth volume was by Mello Leitão). After Oswaldo Aranha accepted the post of minister of foreign affairs in 1938, he had worked relentlessly to enhance Pan-American relations and had signed agreements with Argentina, Chile, Uruguay, Paraguay, Colombia, Panama, the Dominican Republic, Venezuela, Bolivia, Cuba, and Peru. In 1939, Brazil participated in the First American Congress of National Commissions for Intellectual Cooperation, held in Chile, where it was represented by the diplomats Abelardo do Prado and Luz Pinto as well as by Roquette-Pinto, then director of the INCE.[42]

In the case of Uruguay, exchanges had begun quite early, with Mello Leitão's 1931 trip kicking off a series of initiatives. The Uruguay-Brazil Institute was founded in Montevideo in August 1940; it had a library and offered courses in Portuguese. The endeavor proved so successful that in July 1945 Uruguayan lyceums began offering Portuguese as an elective,[43] just a few short months before Mello Leitão arrived there in the company of the Brazilian mission.

Mello Leitão was quite familiar with the spiders of Uruguay and Argentina. When he arrived in these countries as an envoy of the Brazilian government, he was immediately received by entomological research institutes, whether his journey was official (as in Montevideo) or not (as in Buenos Aires and La Plata). He published in local journals, visited museums and got to know their collections, met with local scientists, and was honored. Through his correspondence and publications, Mello Leitão became part of a broad Pan-American network devoted to arachnids, and through his work he in turn reinforced this web. While Oswaldo Aranha was busy signing cultural agreements until his resignation in 1944, Mello Leitão exchanged spiders, scorpions, and offprints with colleagues in a wide range of countries.

A perusal of some brief introductory notes to his papers provides a notion of this give-and-take. In 1931, Mello Leitão published a paper on the arachnids of Argentina in which he expressed his gratitude to Emilio Gemignani of the Bernardino Rivadávia Museum of Natural History in Buenos Aires and to Carlos Bruch and José Canals, both of the La Plata Museum of Natural Sciences, all of whom had sent him specimens. In another paper, published in 1939, he thanked "Mr. Richards," of the Imperial College of Science and Technology in London, for shipping him a small collection of spiders collected in British Guiana. This contact would be repeated in 1941. Shortly before he passed away, Mello Leitão published a paper on three hundred spiders of British Guiana; the specimens had been sent to him by Britain's Natural History Museum, thanks to the kindness of the researcher E. Browning. There are myriad similar examples from other countries and from regions in Brazil. This network really began flourishing in 1940, as Mello Leitão gained surer footing as an internationally respected arachnologist. In publications from 1940 on, he conveys his gratitude to various researchers, including Ergasto Cordero, for daddy longlegs collected around Caracas; Professor Lizer y Trelles, from Buenos Aires, for specimens gathered in the vicinity of Mount Aconcagua; and Ruiz Leal, for daddy longlegs from different locations in Colombia. The U.S. arachnologist Harriet Exline Frizzell had kindly sent him spiders from different places in Peru, Mexico, and Ecuador, and he had identified and described several new species among them. A naturalist priest in Colombia, Nicéforo María, had sent Mello Leitão a sizable collection of Colombian arachnids, which enabled him to assemble an early catalog of spiders from Colombia, where very little was known about the country's arachnid population. Francisco Campos and Carlos Porter, professors in Ecuador and Chile, respectively, had shipped him spiders in 1945. One of his last publications—posthumous in fact—acknowledges "Sr. W. Weyrauch," of Peru, who provided a small collection of arachnids gathered in different parts of the country.[44]

The periodicals to which Mello Leitão submitted his work and the different languages in which they were published also speak to his success within this worldwide web of researchers. Of his 212 papers on zoology (198 of which are about arachnids), 69 appeared in foreign journals and 61 in other languages (French, Spanish, or English).[45] With each publication, he sent offprints to a roster of colleagues in other countries, a fact corroborated in his correspondence with the National Museum after he left it and in the archives of natural history museum libraries in Montevideo, La Plata, and Buenos Aires. There are 41 offprints of papers by Mello Leitão at the library of the National Museum of Natural History in Montevideo, many stamped with the words "compliments of the author." Others bear handwritten dedications, such as "with kindest regards," "with the sincerest friendship of Mello Leitão," or "with fondest remembrance from Mello Leitão." The libraries of the La Plata Museum of Natural Sciences and the Bernardino Rivadávia Museum of Natural History in Buenos Aires hold some offprints, but they mostly have copies of his books in Portuguese, like *A biologia no Brasil*, *Zoogeografia do Brasil*, and *História das expedições científicas no Brasil*. Since Mello Leitão forwarded copies of his writings to other researchers, we can surmise that many other articles and books could be found in private archives or in the individual files of scientists in institutional archives.

SOUTH AMERICAN SPIDERS

Biogeography—a branch of biology devoted to the distribution of living organisms across space and time—surpassed mere collection building, natural history, and morphology and instead built knowledge at the crossroads of various fields, such as geography, climatology, geology, ecology, evolution, and paleontology. The field was born within the heart of natural history during the eighteenth century, first with the writings of Linnaeus and Buffon—often called the "father" of biogeography by the history of science—and, in the nineteenth century, with the work of Humboldt and Louis Agassiz. In 1859, when Darwin devoted chapters 11 and 12 of *On the Origin of Species* to biogeography, he radically revised the field, shifting it away from creationist explanations and natural history toward the study of populations of living organisms. At almost the same time, Alfred Russel Wallace (1823–1913) laid the foundations for new inquiries in the manuscript he delivered to Darwin in 1858, following his scientific observations on the Malay Archipelago.[46] Despite Darwin's contribution to a renewal of biogeography, the theoretical bases of the discipline followed the prevailing currents among biologists during the early decades of the twentieth century, that is, rejection of the Darwinist theory of evolution based on natural selection. In the early 1930s, Mello Leitão dedicated a number of studies to zoogeography from an evolutionary perspective based on the idea of harmonious, cooperative relations between living organisms and drawing from the theory of mutual aid.

Because of his biogeographical perspective, Mello Leitão's career far exceeded simple engagement with other arachnid scholars. As we have seen, he published journal papers on species in Argentina, Colombia, Paraguay, and Peru, among other countries, and became a leading figure in the study of South American spiders. He never confined himself to national borders but selected the continent as his geographical unit, based on notions of a South American zoogeography. Rejecting the pigeonholes of "Brazilian spiders" or "Argentinean spiders," he focused on identifying the occurrence of South American species. As he saw it, given the contiguity of Latin American borders and the similarity of physical and climatic conditions, boundaries between countries were often useless abstractions when it came to understanding spiders.

Mello Leitão's travels to Uruguay as a Brazilian cultural representative in 1931 and 1945 laid the groundwork for a number of important transformations. Although some things remained constant—for example, by 1931 he had already identified himself as a zoologist and scholar of living organisms and their biogeography, lecturing on the "wonderful life of spiders" and the "zoogeography of South American daddy longlegs"—many other things changed over this period.

In 1931, the newly instated Provisional Government was stirring great expectations in Brazil. Hopes were high that a harmonious, corporatist, conflict-free society could be built, ruled by a strong, central state under the firm leadership of Getúlio Vargas. All of this, according to the new line of thought, contrasted with the exaggerated decentralization of the republic before 1930, which had been governed by selfishness, individual interests, competition, and the victory of the strong over the weak and unprotected. This was not an isolated movement, for liberal values were in sharp decline worldwide.[47] The profound rejection of Darwin then prevalent among scholars of biology had political connotations and should be understood in this larger historical framework, as I argued in the first chapter.

When Mello Leitão took his trip in 1945, however, things were quite different. The final years of the Estado Novo had seen a major backlash in Brazilian civil society. The press and public opinion persistently criticized the government for contradictorily backing democratic regimes abroad while maintaining a dictatorship at home. In October 1943, the Mineiros Manifesto, signed by influential members of the liberal elite, demanded a return to democracy. The same year, Brazil's powerful National Student Union organized an important drive against authoritarianism, culminating with a march in December. Oswaldo Aranha resigned as minister of foreign affairs in 1944 in protest against the government's repressive actions.[48] In the international arena, the United States emerged as the paramount defender of world freedom, and Western societies looked upon political and economic liberalism with increasingly kinder eyes.

It was within this context that Mello Leitão introduced new elements into his biological perspective. Even though it cannot be said that he adopted a radical change of paradigm, it is interesting to note a softening of his rejection of Darwinism. In the paper he presented in Montevideo, entitled "New Directions in Biogeography" (delivered in Spanish but published in Portuguese in 1945), he makes observations about variation and natural selection, citing both Darwin and the ornithologist Ernst Mayr, one of the celebrated biologists responsible for the rehabilitation of Darwinism. In his text, Mello Leitão specifically references the "very recent definition of species" by Mayr. In a particular item on selection, he explains that, given the pressure caused by the variability of living organisms, "new forms and competition surface within one species, triggered precisely by excessive density and an inadequate food supply." A few lines later, he refers to the deep impression that the distribution of South American animals had made on Darwin and the importance accorded to geographic isolation in the theory of the origin of species.[49]

Between 1931 and 1945, Mello Leitão forged ties with Latin America at the same time that he was studying incidences of arachnids in his South American zoogeography. In 1931, in preambles to his lectures, he rued the fact that he did not know Spanish and said he hoped he would be understood if he spoke slowly. In an effort to overcome the language barrier, he relied on a resource that had become routine at the National Museum and illustrated his presentation with a slide show. His apology was sincere: when Uruguayan newspapers announced his lectures in 1945, they emphasized that the arachnologist would present all of his talks in Spanish.[50]

The idea of a Pan-American community of entomologists also comes across quite clearly in the 1946 release of Mello Leitão's *Glossário biológico*, a general reference work of technical terms. He dedicated the volume to professors Angelo Moreira da Costa Lima of the National School of Agronomy of Rio de

Janeiro, Ergasto Cordero of the National Museum of Natural History in Montevideo, and Max Biraben of the National University of La Plata, models of "wise men who are an honor to South American science."[51]

LIFE ON EARTH

In 1947, when Mello Leitão published the second edition of *Zoogeografia do Brasil*, he said it was practically a new book, both because it was so much longer but also because its approach was so different. The author had caught sight of a reality that showed no respect for the borders on political maps, and his new knowledge had tempered the nationalist leanings that had originally spurred his sharp interest in fauna inside Brazil.

Looking beyond Mello Leitão's experiences in scientific exchange, we can discern something else that accounts for his somewhat conflicted nationalist leanings, something directly bound up with the history of science in his day. Biogeography was still a recent field then, receiving a great impetus in the twentieth century as it interacted with the dilemmas and challenges of a rapidly changing world. As Mello Leitão himself noted, better means of communication and safer, faster travel had facilitated, as never before, access to places where people could find forms of life distinct from those in their native lands, pressing home the actuality of biological diversity. These increasingly frequent encounters with the diversity of life deepened people's perception of world space. Biogeographical research boosted knowledge about the long history of planetwide movements of diverse plant and animal species. If nineteenth-century European naturalists visiting foreign shores had shipped taxidermied specimens of exotic fauna back to the shelves of Old World museums, twentieth-century biologists shifted their gaze to the voluntary or passive flows of life. The idea of people in motion was joined by the notion of a planet of living organisms taking part in extended, complex spatial dynamics.[52]

Mello Leitão's contemporaries discovered the instability of what once seemed stable. The very ground over which humans, animals, and plants moved began to move as well. The classic theory of continental bridges—according to which the original connections between permanent continental blocks had eventually worn away or sunk—was proving ever more unsatisfactory. In 1915, Alfred Wegener proposed his continental drift theory. Now part of our worldview, the theory was at first harshly repudiated in scientific circles. Although

Wegener's book was republished in 1920, 1922, and 1929, his ideas only came to be accepted in the late 1950s, when studies of paleomagnetism furnished a new understanding of the marks that the earth's magnetic field had left on rocks, and it thus became possible to draw inferences about the behavior of the magnetic field in earlier times and about the shifting of tectonic plates.[53]

Our Brazilian zoologist was quick to warm to the theory of continental drift, beguiled by the idea that the continental masses had once made up a common land, Pangaea, but had then moved apart. It seemed to be a persuasive explanation for the history of the distribution of living organisms across the earth's surface. On a shared planet, the risks of extinction were a warning to the world. By consulting ornithological checklists, Mello Leitão tallied the birds that had recently disappeared from different corners of the world. He also listed the dangers of indiscriminate hunting for food, adornment, or sporting pleasure. He noted the destruction of habitats through agricultural activity and pollution as well as through the imbalances resulting from people's careless management of species. He pointed to species that had been involuntarily dispersed by the action of man—like snails, fleas, cockroaches, fungi, mosquitoes, and parasites—ultimately harming human societies. He described imbalances and extinctions taking place across all continents and oceans. Man was at fault, because he dispersed species unsustainably, disturbed precious areas of isolation, and destroyed natural habitats, thereby recklessly disrupting the unity and diversity of life on earth, with unforeseeable consequences. Alongside processes of extinction, environmental catastrophes were already affecting the world in those years, an issue that went well beyond national borders. Mello Leitão cites an international committee to defend fauna that denounced the death of massive numbers of aquatic birds after ships discharged spent crude oil into the ocean. Even though Pangaea had been split apart, the bonds between life on earth would not be broken without consequences.[54] As a man of his time, Mello Leitão was influenced by a kind of global perception, which now, decades later, is one of the most notable features of our contemporary thinking about nature.

SCHOLAR AND EDUCATOR

On May 11, 1943, Mello Leitão attended a special meeting of the ABC, of which he had been a member since 1917. It was his swearing-in ceremony as the academy's new president. He was replacing Arthur Moses (1886–1967)[55]—his friend since medical school, a biologist at the Oswaldo Cruz Institute, and the speaker who would welcome him as president.

Those were somber times. The previous summer, Germany had torpedoed a series of Brazilian ships, compelling the Brazilian government to declare war on August 31. In a meeting held just after the attacks, the academy approved a motion in support of the government's decision. When Mello Leitão took the floor to defend the motion, he declared that all Brazilians were outraged and in mourning. A few months later, during his inaugural speech at the academy, he hailed Brazil's decision to send troops to Italy and stressed that scientists could make a crucial contribution to the war effort. Moreover, he was hopeful about the eventual "victory of Civilization."[56]

When Arthur Moses introduced Mello Leitão as the next president, he listed the considerations that ranked him as an exceptional scientist. In addition to his prodigious specialized scholarship, Moses said, Mello Leitão had written outstanding textbooks for high school and college students that introduced state-of-the-art information and relied predominantly on examples of Brazilian flora and fauna. Moses also noted that the Brazilian arachnologist's body of scientific work was well known throughout the international community, and he cited the illustrious entomologists who had sung his colleague's praises: Clarence Hoffman, for example, had labeled him the world's foremost expert on South American spiders, while Alexander Petrunkevitch declared his treatises to be seminal works in the field.[57] Moses praised Mello Leitão as an active, steadfast participant at ABC meetings, where he joined in discussions and presented papers.

Since its founding in 1916, the academy had promoted scientific specialization and the formation of distinct fields of knowledge.[58] During his inaugural speech in 1943, Mello Leitão spoke about the construction of the field of biology in Brazil, focusing specifically on zoology. He said zoologists in the twentieth century were playing an increasingly active role in society. They had pointed out the distinction between *Aedes aegypti* and other, harmless mosquitoes, after researching and demonstrating their biological cycle, habits, and ecology, making it possible to fight yellow fever in major urban areas throughout Brazil. By studying each region's fauna, zoologists had been able to assess the real risks of the occurrence of certain human illnesses. In addition to associating the presence of *Aedes aegypti* with yellow fever, they had also linked *anophelines* to malaria, sand flies to leishmaniosis, triatomes (kissing bugs) to Chagas disease, and *Australorbis* to South American schistosomiasis. Economically speaking, the work of zoologists had also proven vital by identifying the causes of the insect plagues that decimated crops and caused animal epidemics. Similarly, the field had drawn a distinction between truly poisonous animals and those that are not only harmless but actually essential to ecological balance in different biogeographic regions. By revealing the life cycles of precious species and indicating the practices most suitable for protecting marine and lacustrine fauna, zoologists had helped improve fishing activities. Moreover, they had shown that much wildlife is extremely useful, warned governments about the risks of extinction, and worked for the creation of reserves and national parks. In short, Mello Leitão argued that zoology lies at the core of human life and the health of populations; is essential to the agricultural, livestock-raising, and fishing industries; and stands as a major force behind the protection of large natural areas—it is a science of and for life.[59] He also asserted that the field performed an economic, social, and, chiefly, political role. Yet after trumpeting biology as a science capable of forecasting with "the rigor of mathematical formulas," exact and unbiased, Mello Leitão slipped up and remarked that biology had cast aside contemplation in favor of "activism" in the twentieth century—giving us evidence of the historical evolution of the field's scientific practices.[60]

Mello Leitão was inarguably a leader in the construction of zoology as a field in Brazil. His 1937 book *A biologia no Brasil* identified the Brazilian pioneers in biology, searched for the field's origins, and offered contemporaries a linear, progressive, and cumulative vision of the history of this science. It was as a zoologist that Mello Leitão joined the University of Brazil in 1939. In late 1937, when the Estado Novo barred people from holding more than one public job, he had left the National Museum and opted for his post at the Higher School of Agricultural Science and Veterinary Medicine. Only two years later, when he had to resign from the Higher School in order to accept a university teaching position, he sent the director an official letter in which he displayed his love for the school where he had spent twenty-six years, arguing that he was merely doing his duty by leaving it, "obeying orders from the President of the Republic to hold the chair of Zoology at the newly established National School of Philosophy."[61] His words show that he felt he had been summoned to take up an important national duty.

No less important was Mello Leitão's defense of the scientific specificity of biological knowledge in his textbooks, used by generations of high school and normal school students. From 1917 to 1946, Mello Leitão published ten textbooks, some of which were multivolume works or appeared in more than one edition. As mentioned earlier, he also released a glossary of scientific terms meant as a reference work for anyone interested in learning biology.[62] Considering that many schools, like the prestigious Colégio Pedro II, continued to call the class Natural History until the 1970s, it is no minor detail that normal schools in Rio de Janeiro and São Paulo added Educational Biology and General Biology to their curricula in 1930.[63]

It is also important to bear in mind the historical nature of school curricula and the fact that syllabus content is not a given; rather, it is the result of complex processes of transformations in disciplines that occur both during social disputes within the greater social framework and also during internal disputes at schools themselves.[64] With this in mind, we should take note that "zoology," "botany," and "biology" were predominant terms in the titles of Mello Leitão's textbooks, supplanting so-called natural history and manifesting his firm conviction that biology—the science of living organisms—should be a specific topic in school curricula. Of his eleven textbooks, only one retained the term "natural history" in its title.

In his preface to *Biologia geral* in 1940, Mello Leitão spelled out his intent to produce textbooks that were in line with the notion of biology as a specialization. Casting aside a tradition confined to morphological descriptions, he wanted this book to afford students a "synthetic view of living organisms and their relations to each other and to the environment in which they live." He accentuated the value of learning the principles of biophysics and biochemistry, cell physiology, reproduction, genetics, and ecology and the critical study of doctrines of evolution—precisely the subjects of his chapters. Even though school curricula still termed the course "natural history," the approach in Mello Leitão's books and his dominant epistemological perspective prodded content toward the teaching of biology.

Mello Leitão's emphasis on images and examples of Brazilian flora and fauna meshed well with nationalist projects, with his advocacy of science communication, and with the goal of expanding knowledge through practical teaching tied to daily life. Textbooks were a strategic element in the construction of biological knowledge in Brazil, and Mello Leitão was one of the most important textbook writers of his day.

Scholars competed not only in authoring textbooks but also in getting them published and adopted at schools. One of these disputes constituted a significant episode in Mello Leitão's career. In a critical review that he submitted at the request of the National Textbook Commission in 1941, he condemned the adoption of the book *Zoologia*, by Waldemiro Potsch, a teacher at Colégio Pedro II. Potsch sued him in 1944, claiming that Mello Leitão's negative review was nothing more than an effort at monopolizing the market in biology books. Besides filing suit, Potsch wrote a lengthy dossier to sustain his accusation that Mello Leitão had committed plagiarism multiple times over in his books. The arachnologist responded by publishing a book in which he refuted the accusations point by point. A report by a court-ordered expert proclaimed his innocence, but the suit dragged on until January 1948, when it was finally decided in his favor, a few months before his death.[65] Mello Leitão was much aggrieved by the lawsuit and the accusations, as is apparent in his autobiography, which closes with a reference to the court decision and the fact that he was cleared of all accusations. Some of his texts from the time of the lawsuit vehemently censure the practice of slander, including the speech he delivered in Montevideo during the 1945 cultural mission: after explaining how man had proven ever more capable of defending himself from the poisons emitted by other animals, Mello Leitão pointed out that there were much more vicious venoms, like slander, which he compared to totalitarianism.[66]

It was in the midst of this dispute that Mello Leitão, already a zoologist of renown, chaired the ABC from 1943 to 1945. Under his presidency, the ABC commemorated the fourth centennial of the death of Copernicus (1473–1543), the bicentennials of the births of Condorcet (1743–94) and Lavoisier (1743–94), and the bicentennial of La Condamine's expedition. Mello Leitão also received Argentinean, U.S., French, and Polish scientists, including Theodosius Dobzhansky, the noteworthy neo-Darwinist geneticist.[67]

FONTAINEBLEAU, 1948

Shortly after the end of World War II, organizations for the protection of nature in a number of countries joined forces to form an international body, following up on a proposal first made over thirty years earlier during the 1913 International Conference on the Protection of Nature in Berne and ratified during a second gathering, held in Paris in 1923. The International Office for the Protection of Nature was subsequently founded in Brussels in 1928, but it grappled with growing challenges due to new global upheaval. After World War II, the Swiss League for the Protection of Nature assumed leadership in a new push to establish a worldwide organization. In 1947, the International

Conference for the Protection of Nature, held in Brunnen, Switzerland, finally approved the formation of an international union. Its members then asked UNESCO to throw its support behind an event in Paris, where the new institution would be founded and organized.[68]

The Conference for the Establishment of the International Union for the Protection of Nature took place in Fontainebleau, France, from September 30 to October 7, 1948. Documents from the event indicate that only one Brazilian was in attendance, Cândido de Mello Leitão, as a representative of the Brazilian government and of the National Commission for the Protection of South American Fauna.[69] It is noteworthy that documentary files held by the National Museum and the ABC contain data on the zoologist's involvement in a wide diversity of organizations but not one word about this nongovernmental commission, which presented itself as Brazilian (it was, after all, a "national" commission) but also proposed to work beyond the borders of the country, targeting the "defense of South American fauna."

Mello Leitão was hardly on the sidelines at the event. He was a member of the steering and program committee and of the finance committee. During the session on the definition of national parks, he took the floor to endorse the adoption of more accessible nomenclatures that would make it easier for ordinary people to understand notions of ecology. His rejection of hermetic language likely to confine the debate to strictly scientific circles was a position eventually embraced by the majority of those present.[70]

The Fontainebleau conference was Mello Leitão's last major activity. He passed away a few weeks later, on December 14, 1948, the victim of cancer. His obituaries, too, ignore his participation at Fontainebleau, obscuring Brazil's (admittedly minimal) participation in the international postwar movement to protect nature.

Over the course of his lengthy professional life, Mello Leitão was more than just a naturalist or even an eccentric pediatrician devoted to collecting and classifying spiders. Little by little, he moved beyond the epistemological limits of natural history, transforming himself into a biologist—more specifically, into an arachnologist. He researched spiders as living organisms and studied their historical distribution across space, investigating not only their anatomy, histology, and physiology but also their ecology and etiology. On top of identifying, classifying, and naming arachnid species, Mello Leitão examined their venoms, habits, and traits: how they built their webs and cared for their offspring; the dispersion of families of cellar spiders; how the structures and arrangements of their eyes influenced their visual acuity and thus important aspects of their lives, like stalking prey. His career began with experiences in collecting and classification but then shifted to the construction of a research perspective of broader scope, in which arachnids were living organisms to be studied, along with their zoogeography. His scientific trajectory stands as an undeniable example of the making of a biologist in Brazil in the first half of the twentieth century.

While bearing in mind that the story of any given individual transpires within a framework of complex social relations and of unfolding history, we must also take into account the coming into being of the historical subject, during which he, along with his contemporaries, devises new practices within his own time and together with his contemporaries. A scholar of spiders, Mello Leitão was indeed an expert in webs, and it was as part of them that he learned his craft. Within these networks of personal and institutional relations, historical actors built and shared knowledge while they constructed roles for themselves and their peers, the institutions where they worked, and the science they practiced.

CONCLUSION

IN JANUARY 2015—summer vacation and high tourist season in Brazil—patrons found the doors of the National Museum closed. The museum usually receives some seven thousand visitors a week during this time of year, but its funds had been abruptly cut off as a consequence of a nationwide political and economic crisis. There was no money to pay for either cleaning or security services, and exhibits were suspended for several weeks. In a widely distributed press release explaining the situation, the museum board decried the fact that such was the fate of Brazil's oldest science institution—just three short years from commemorating its bicentennial, in 2018.

Today, the National Museum is part of the Federal University of Rio de Janeiro, in turn under the aegis of the Ministry of Education. Nestled in the midst of nearly eleven acres of botanical gardens, the museum is home to historical archives, libraries, research laboratories, and graduate courses, as well as host to public exhibits on biological anthropology, archaeology, ethnology, geology, paleontology, and zoology. Since 1927, one of its extension projects, the Assistance Service for the Teaching of Natural History, has been designing and sponsoring educational initiatives, advising teachers, and stimulating the minds of young students.

The National Museum has long been a privileged locus for scholarly reflection on Brazil. In the nineteenth century, it sheltered the dreams of the newly independent nation's intellectual elites, who wanted to join Europe as part of the civilized world. In the early twentieth century, the museum served as a point of departure for the conquest and rediscovery of Brazil, as its members turned their attention to the interior of the nation and its sertões and thick forests. Flora and fauna, minerals, indigenes, and people of mixed descent were viewed as hieroglyphics waiting to be deciphered in the hope that they would reveal the path to nation building. The National Museum's current mission is to be a steward of Brazil's scientific memory and produce innovative knowledge applicable to environmental protection and social transformation. Its limitations and accomplishments mirror the educational, cultural, and research challenges peculiar to Brazil. When I heard it had closed its doors for lack of funding, I found myself asking how Edgard Roquette-Pinto, Alberto José de Sampaio, and Cândido de Mello Leitão would react to the news if they were alive today and could see the museum continuing on shaky ground, vulnerable to the prevailing political and economic winds—a status quo reminiscent of their times and always the source of great consternation.

This book has explored the history of the National Museum from 1926 to 1945, focusing on the union of scientific practice and political life, the emergence of scientific specializations and initiatives in science communication, and the careers of three of the museum's most active members. In Brazil, the time frame of this study witnessed the formation of biology as a field in its own right—a historical transformation all the more meaningful because many of the activities taken up by this nascent science displayed an inherent political content. The excitement and creativity of these researchers as they set about modeling new ways of producing and communicating knowledge must be understood in the broader context of the political battles and social confrontations then underway in Brazilian society. Yet much more than simply echoing or stemming from a particular context, the scientific practices that were forged in the hallways, laboratories, and workshops of the National Museum were an integral part of the making of history during those years.

Writing at the juncture of political history and the history of science, I have tried to steer away from any dualistic criticism of these scientists. A critique of reason as the only possible form of knowledge should not lead us to discredit reason out of hand, for that would bear the mark of intolerance. Scientists are condemned neither to authoritarian postures nor to blind rationalism. Their relations with the powers that be do not always put them in conflict with a population that is the victim of their attentions and interventions. When science and power work hand in hand, the purpose is not necessarily to refine new strategies for domination and social heteronomy. Knowledge in any form—be it scientific, folk, traditional, artistic, or historical—involves the exercise of power, and the relations between science and power are extremely complex. Historiography must shun its penchant for assigning the role of victim to some and oppressor to others. When we demonstrate that the historical viewpoint is vital to the study of scientific practices, we lend intelligibility to scientific endeavor, and this should be one of the overriding goals of the history of science.[1] By examining the history of the production of knowledge, we are able to take into account the larger background of the political, cultural, and social circumstances in which scientists have engendered and shared new knowledge. If we are to arrive at a more rational understanding of the role that scientific endeavor has played in the formation of societies down through time, we must avoid what Marc Bloch called the "satanic enemy of true history: the mania for making judgments."[2]

Nor can the history of science be written as a saga of pioneers, which would be the flip side of taking a dualistic attitude. As I wrote in the introduction, it might be tempting to interpret the work done at the National Museum as indicative of the awakening of many of our current apprehensions in the realm of environmental initiatives, inclusive education, and myriad other areas. The search for origins tends to blind the historian to the complexity of historical development, obscuring the manifold possibilities that were at play in the past as well as the fact that, at any given moment, the future is indefinite. When impelled by an "embryogenic obsession"—to borrow Bloch's term—we run a great risk not just of searching for an origin that explains but also of believing that this beginning might suffice as an explanation.[3]

When I researched the professional histories of Roquette-Pinto, Mello Leitão, and Sampaio, I observed how they had shaped their scientific activities at a time when various fields of knowledge were still blurred with natural history and how this enabled them to transit between disciplines and among scientific, artistic, and technological projects. This same set of circumstances also enriched the science communication efforts that they embraced with such idealistic enthusiasm. Yet, all the while, they pursued specialization in their own fields. The most successful example in this regard was Mello Leitão, who became a nationally and internationally recognized biologist. As these men fought to promote a renaissance at the National Museum, they also wanted the institution to foster specialized production in a range of areas, with biology serving as a flagship of the sciences. They presented the field as key to addressing and potentially solving many of the problems of life that were then having an impact on human, animal, and plant populations all across the national territory. At the same time, the reigning attitude was authoritarian; in the eyes of these National Museum scientists, the only truly valid knowledge was the knowledge they possessed and wanted to disseminate. Their voluntarism derived from the belief that historical change should be accomplished by an intellectual elite rather than achieved through the effective construction of citizenship.

These men were intensely involved with the society in which they lived; they were committed to debating problems and solutions, devising strategies, and taking part in the game of politics while trying to put politics at the service of their proposals and their dreams. Their actions, expectations, and criticisms were imbued with a boldness and inventiveness that inspires reflection, as does the backdrop of turmoil and paradoxes at the National Museum in their time. For all these reasons, the history of Brazil's activist biology poses the intellectual enterprise as one of creative action and the construction of knowledge as a form of potentially transformative, engagé political praxis.

TIMELINE OF BRAZILIAN HISTORY (1889–1945)

FIRST REPUBLIC

1889–1930 Brazil declared its independence from Portugal in 1822 but maintained the monarchy, under the rule of the Bragança dynasty. Slavery lasted until 1888. On November 15, 1889, members of the army established the republic. The Constitution of 1891 stipulated that the new republican state would be democratic, presidentialist, federalist, and secular. The military dominated the first two presidential administrations, but as of 1894, civilian political groups held sway. State oligarchies with ties to commodity production, especially coffee, relied on election fraud to control the presidency and Congress. Because illiterate persons were denied the right to vote, much of the population was excluded from formal political life, and social inequality remained strong. Although this period of Brazilian history is often viewed negatively—and pejoratively called the Old Republic—it was also a time of industrialization, urban growth, immigration, the formation of a working class and labor movement, the rise of cultural and intellectual movements, and the firm demarcation of the nation's borders.

REVOLUTION OF 1930

1930 This armed movement began on October 3, 1930, when discontented dissident oligarchs joined forces with pockets of the middle classes and with young army officers (the latter known as *tenentismo*). All were critical of what they saw as the excessive liberalism of the First Republic, corrupt elections, and the government's weak response to the country's problems. Victorious on November 3, 1930, the movement carried Getúlio Vargas into power and put in place a nationalist, centralizing, and interventionist state. It was a watershed in the history of the Brazilian republic, inaugurating economic nationalism under the strong arm of the state. Historians have long debated whether the movement should in fact be labeled a "revolution." Some contend it was just the opposite: a strategic rearrangement by the ruling elites meant to contain the leftist revolutionary proposals espoused both by anarchist unions, since the 1910s, and by the Brazilian Communist Party, founded in 1922.

PROVISIONAL GOVERNMENT

1930–1934 Under the Provisional Government, Vargas ruled by so-called decree-laws. The new government comprised a number of groups, with young army officers predominating. The latter advocated a strong, centralized government, statism, economic diversification, infrastructure development, and the enactment of new health and education policies, while they also wanted social rights to remain under state control. In order to facilitate implementation of these measures, elections to the Constituent Assembly were postponed. Significant changes took place from 1930 to 1934: the enactment of labor laws, public education reform, the creation of both electoral and labor tribunals, passage of the secret ballot, and granting women the right to vote. The government also decreed a number of codes in regard to nature and territory, like the Game and Fish Code, Forest Code, Mine Code, Water Code, and Animal

Protection Code. The constitution was finally enacted on July 16, 1934, and Congress elected Vargas president. New, direct elections were slated to take place in 1938, when Vargas would not be eligible to run again. The fact that conservative Catholic groups had been gaining ground was embedded in the preamble to the new constitution, which invoked the name of God, countering the lay nature of the Brazilian state as established following the Proclamation of the Republic.

CONSTITUTIONAL GOVERNMENT

1934–1937 When Vargas took office as president, he was unhappy about the new constitution, which limited his powers, which until then had been wholly discretionary. To stand strong against the political class, he forged tight bonds with the army's high command. In a complex political situation, social conflicts and right- and left-wing political radicalism made inroads. In 1935, a number of political sectors joined forces to fight fascism and imperialism by founding the National Liberation Alliance, drawing thousands of supporters across the country. That same year saw a wave of strikes, and these, plus the alleged Communist threat, were invoked to justify enactment by decree of the National Security Law, which abolished democratic guarantees and defined crimes against the state. Escalating political clashes and a climate of anticommunist paranoia culminated in the November 1937 military coup. With the support of the high command of the armed forces, Vargas shut down Congress and transformed his presidency into a dictatorship.

ESTADO NOVO

1937–1945 In 1937, Getúlio Vargas enacted an authoritarian, centralizing constitution that placed great power in the hands of the head of the executive branch. Government intervention in the economy was the order of the day, aimed at fostering development and

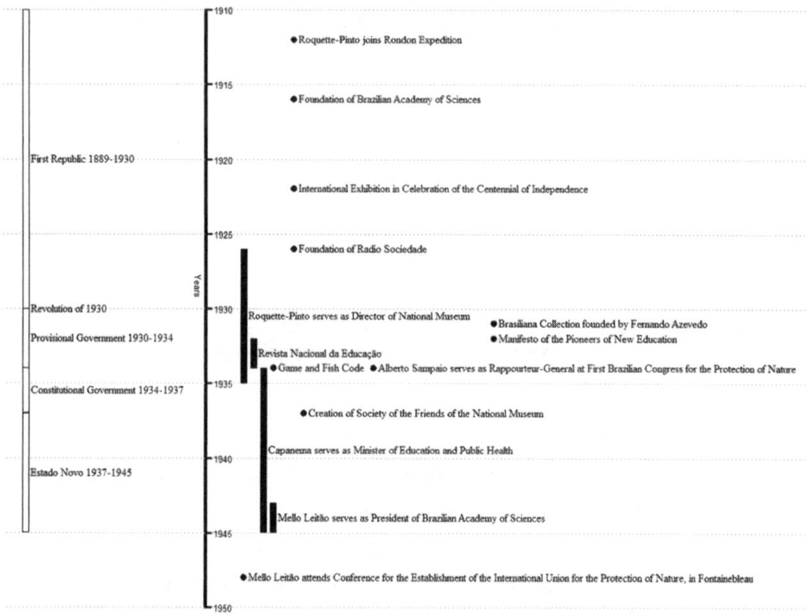

FIGURE 26. Timeline. Compiled by Regina Horta Duarte.

industrialization. Vargas intended to shape a new nation and regenerate the lives of Brazilians. In 1939, he founded the Press and Propaganda Department, an agency that engaged in intense, systematic political propaganda through all media and was also responsible for censoring any cultural or intellectual expression against the regime. The Estado Novo was critical of Brazil's liberal democratic past, which it associated with the corruption characteristic of the early years of the republic. In April 1942, the country joined the fight against the Axis alliance, acting in blatant contrast to the authoritarian, corporatist nature of the Vargas regime. When World War II ended, the victory of liberal democracy in the West shook the foundations of the Estado Novo. On October 29, 1945, Vargas was deposed by the army. Since there was no vice president, the Federal Supreme Court held power until elections took place.

NOTES

INTRODUCTION

1. On the Portuguese court's move to Rio de Janeiro and the ensuing transformations in Brazil, see Maria Odila Leite Dias, "A interiorização da metrópole," in *1822: Dimensões*, ed. Carlos Guilherme Motta, 160–84 (São Paulo: Perspectiva, 1972); Maria de Lourdes Viana Lyra, *A utopia do poderoso império* (Rio de Janeiro: Sette Letras, 1994), 131–72; Jurandir Malerba, *A Corte no exílio* (São Paulo: Companhia das Letras, 2001). On the establishment of the National Museum in Rio de Janeiro in 1818, see Margareth Lopes, *O Brasil descobre a pesquisa científica* (São Paulo: Hucitec, 1997), 25–70; Silvia Figueiroa, "Mundialização da ciência e respostas locais," *Asclepio* 50:2 (1998): 107–23.
2. Susan Sheets-Pyenson, *Cathedrals of Science: The Development of Colonial Natural History Museums during the Late Nineteenth Century* (Kingston: McGill-Queen's University Press, 1988), 4. On the establishment of museums in Latin America, see also Margareth Lopes, "A formação dos museus nacionais e a América Latina independente," *Anais do Museu Histórico Nacional* 30:1 (1998): 121–33; Thomas Glick, "Science and Independence in Latin America," *Hispanic American Historical Review* 71:2 (1991): 306–34.
3. Nancy Stepan, *Beginnings of Brazilian Science: Oswaldo Cruz, Medical Research and Policy, 1890–1920* (New York: Science History Publications, 1976), 36–39.

4. On the proactive role of Latin American scholars and institutions, see Juan José Saldaña and Consuelo Cuevas, "La invención en Mexico de la investigación científica professional," *Quipu* 12:3 (1999): 309–22; Mercedez Valero and Armando García, "Ciencia y colecionismo en Cuba en el siglo XIX," *Asclepio* 51:1 (1999): 205–26; Irina Podgorny, "El ordenamiento de la naturaleza a través de los museos y de la ciencia en la Confederacíon Argentina," *Quipu* 12:2 (1999): 167–86; Margareth Lopes, "Cooperacão científica na América latina," *Interciencias* 25:5 (2000): 228–33; Horácio Capel, "El asociacionismo científico en Iberoamerica," *Interciencias* 17:3 (1992): 168–76; Pedro Pruna, "National Science in a Colonial Context," *Isis* 85:3 (1994): 412–26; Regina Horta Duarte, "Between the National and the Universal: Natural History Networks in Latin American in the Nineteenth and Twentieth Centuries," *Isis* 104:4 (2013): 777–87.
5. Marcus Vinicius de Freitas, *Charles Frederick Hartt, um naturalista no império de Pedro II* (Belo Horizonte: Editora UFMG, 2002); David A. West, *Fritz Müller: A Naturalist in Brazil* (Blacksburg: Pocahontas Press, 2003); Nelson Sanjad, *Emílio Goeldi (1859–1917): A ventura de um naturalista entre a Europa e o* Brasil (Rio de Janeiro: EMC Edições, 2009); Maria do Carmo Gomes, *A pequena pátria: Políticas cartográficas em Minas Gerais* (São Paulo: AnnaBlume, 2015).
6. Lopes, *O Brasil descobre a pesquisa*, 158–293.
7. Alda Heizer, "Considerações sobre a participação da América Latina nas grandes Exposições," *Revista da ANPLAC* 9:1 (2010): 15–28; Heloisa Domingues, "As demandas científicas e a participação do Brasil nas Exposições Internacionais," *Quipu* 12:2 (1999): 203–15; Sandra Pesavento, "Imagens da nação, do progresso e da tecnologia: A Exposição Universal de 1876," *Anais do Museu Paulista* 2:1 (1994): 151–67; Lilia Schwarcz, "Os trópicos como espetáculo," in *Galerias del Progreso*, ed. Beatriz González and Jens Andermann, 195–220 (Buenos Aires: Beatriz Viterbo Editora, 2006); Livia Rezende, "Crafting the Nation: Brazilian 'Civilised Exoticism' at World Exhibitions," in *Design and Craft: A History of Convergences and Divergences*, ed. Javier Gimeno and Fredie Floré, 136–39 (Brussels: Universa, 2010); Jens Andermann, "Espetáculos da diferença: A Exposição Antropológica Brasileira de 1882," in *Galerias del progreso*, ed. González and Andermann, 151–93.
8. Jens Andermann, *The Optic of the State* (Pittsburgh: University of Pittsburgh Press, 2007), 1–21.

9. On the oligarchical, exclusionary nature of the republic from 1889 to 1930, see José Murilo de Carvalho, *Os bestializados* (São Paulo: Companhia das Letras, 1987), 44–45; Maria Efigênia Lage de Resende, "O processo político no Primeira República e o liberalismo oligárquico," in *O Brasil Republicano: O tempo do liberalismo excludente*, ed. Jorge Ferreira and Lucília Delgado, 89–120 (Rio de Janeiro: Civilização Brasileira, 2003); Hebe Mattos, "Vida política," in *História do Brasil Nação: A abertura para o mundo*, ed. Lilia Schwarcz, 85–132 (São Paulo: Objetiva, 2013); Francisco Iglésias, *Trajetória política do Brasil* (São Paulo: Companhia das Letras, 1995), 206.
10. On the importance of 1922, see Marieta de Moraes Ferreira, "A reação republicana e a crise política nos anos 1920," *Estudos Historicos* 6:11 (1993): 9–23; Marly Silva da Motta, *A nação faz cem anos* (Rio de Janeiro: CPDOC/FGV, 1992); Livia Rezende, "Designing the State at Brazil's Independence Centennial International Exhibition," in *Design Frontiers*, ed. Priscila Farias and Paul Atkinson, 79–89 (Mexico City: Editorial Designeo, 2014).
11. Peter Hoffenberg, *An Empire on Display* (Berkeley: University of California Press, 2001), xvii.
12. Simon Schwartzman, *Um espaço para a ciência* (Brasília: MCT, 2001), 136–79; Simon Schwartzman, Helena Bomeny, and Vanda Costa, *Tempos de Capanema* (Rio de Janeiro: Paz e Terra, 2000), 221–29.
13. Michel Foucault, *Society Must Be Defended: Lectures at the Collège de France, 1975–1976* (New York: Picador, 2003), 252–54; Foucault, *Security, Territory, Population: Lectures at the Collège de France, 1977–1978* (New York: Picador, 2009), 1.
14. Marc Bloch, *The Historian's Craft* (New York: Vintage, 1964), 29–34.

CHAPTER 1

1. Although Brazil's 1934 Game and Fish Code has been discussed by some authors, they have ignored the draft law. See, e.g., José Augusto Drummond, "A legislação ambiental brasileira de 1934 a 1988," *Ambiente e Sociedade* 2:3–4 (1998/99): 127–49; Carolina Capanema, "A natureza no projeto de construção de um Brasil moderno e a obra de Alberto José de Sampaio" (MA thesis, Universidade Federal de Minas Gerais, Belo Horizonte, 2006), 28; José Luiz de Andrade Franco and José Augusto Drummond, *Proteção à natureza e identidade nacional no Brasil, anos 1920–1940* (Rio de Janeiro: Editora Fiocruz, 2009).

2. Dulce Pandolfi, "Os anos 1930, as incertezas do regime," in *O Brasil Republicano: O tempo do nacional estatismo*, ed. Jorge Ferreira and Lucília Delgado, 13–37 (Rio de Janeiro: Civilização Brasileira, 2003). See also Bóris Fausto, *A Revolução de 30* (São Paulo: Brasiliense, 1970); Edgard De Decca, *1930: O silêncio dos vencidos* (São Paulo: Brasiliense, 1981). On cultural conflicts that began in 1930 over the definition of the essence of Brazilianness, see Daryle Williams, *Culture Wars in Brazil: The First Vargas Regime, 1930–1945* (Durham: Duke University Press, 2001).
3. Mello Leitão, "Protegendo a nossa fauna silvestre (Entrevista)," *A Noite*, February 1933, newspaper clipping, MN.JF.O.MN.DR3, Seção de Memória e Arquivo do Museu Nacional, Universidade Federal do Rio de Janeiro, Brazil (hereafter SEMEAR). On Brazilian naturalists' discontent over the shipment of collections of Brazilian flora and fauna to European museums, beginning in the nineteenth century, see Lopes, *O Brasil descobre a pesquisa científica*, 56, 85.
4. Roquette-Pinto, Sampaio, and Mello Leitão, "Considerações gerais: Projeto de decreto regulamentando a caça silvestre," Anexo ao Ofício 195, October 21, 1932, pp. 7, 22, BR.MN.MN.DR, SEMEAR.
5. Roquette-Pinto, Sampaio, and Leitão, "Considerações gerais," folio 2.
6. "Projeto de Decreto para regulamentação da caça no território nacional, mandado publicar para conhecimento dos cidadãos que queiram apresentar ao Ministro da Educação sugestões a respeito, no prazo de três meses, contados da primeira publicação," *Diário Official da União*, November 22, 1932.
7. According to the preamble submitted to the minister, it would be up to the scientific consultants to decide whether the overpopulation of jaguars might pose a threat to people in the sertões by attacking their cattle; under these circumstances and in these areas, hunting would be permitted until the situation normalized. Roquette-Pinto, Sampaio, and Mello Leitão, "Considerações gerais," 10.
8. Emílio Goeldi, "Destruição das garças e guarás," *Boletim do Museu Paraense* 2 (1898): 27–42; Goeldi, *As aves do Brasil* (Rio de Janeiro: Francisco Alves, 1922), 242–43. On the topic of birds and scientists in Brazil in the early decades of the republic, see Regina Horta Duarte, "Pássaros e cientistas no Brasil: Em busca de proteção, 1894–1938," *Latin American Research Review* 41:1 (2006): 3–26. See also Olivério M. de Oliveira Pinto, "A zoologia no Brasil," in *As ciências no Brasil*, ed. Fernando Azevedo, 2:93–148 (São Paulo: Melhoramentos, 1956). On Goeldi, see Sanjad, *Emílio Goeldi*.
9. Hermann von Ihering, "Necessidade de uma lei federal de caça e proteção das

aves," *Revista do Museu Paulista* 3 (1902): 228–60; "Biografia de Hermann Von Ihering," *Natureza em Revista* 2 (1971): 6–10; Lopes, *O Brasil descobre a pesquisa científica*, 158–212, 248–91.

10. I am drawing from the notion of a "pedagogy of nationality," as developed by Eliana R. de Freitas Dutra, *Rebeldes literários da República: História e identidade nacional no Almanaque Brasileiro Garnier, 1903–1914* (Belo Horizonte: Editora UFMG, 2005), 10.

11. Goeldi, "Destruição das garças," 36; Ihering, "Necessidade de uma lei," 244–49; Ihering, "Proteção às aves," *Revista do Museu Paulista* 9 (1914): 316–32; Ihering, "Devastação e conservação das matas," *Revista do Museu Paulista* 8 (1911): 485–500.

12. Two examples can help put this sum in perspective. In 1931, Roquette-Pinto earned 36 contos de réis a year as director of the National Museum, while Alberto Sampaio and Mello Leitão earned 24 and 19 a year, respectively. See Roquette-Pinto, "Ofício 391, October 15, 1932," RA79.D79, SEMEAR. That same year, the U.S. dollar fell to its lowest level for the entire First Republic, with one mil-réis worth $0.07, meaning that 380 contos de réis equaled $26,600. For exchange rates between the U.S. dollar and the mil-réis from 1880 to 1940, see Thomas Holloway, *Immigrants on the Land: Coffee and Society in São Paulo, 1886–1934* (Chapel Hill: University of North Carolina Press 1980), 181.

13. "Na selva de Matto Grosso, Alexandre Siemel, matador de onças," *A Noite Ilustrada*, January 28, 1931; "Um thesouro a mais, o lagarto e o sapo boio, pelles de ouro, onças, caitetus, giboias e sucuris," *A Noite Ilustrada*, May 20, 1931; "Caçadas e caçadores," *A Noite Ilustrada*, August 14, 1935.

14. Nicolau Sevcenko, "A capital irradiante: Técnica, ritmos e ritos do Rio," in *História da vida privada no Brasil: República, da Belle Époque à Era do Rádio*, ed. Sevcenko, 534–38 (São Paulo: Companhia das Letras, 1998); James Laver, *Costume and Fashion: A Concise History* (Oxford: Oxford University Press, 1983), 216–29; François Boucher, *A History of Costume in the West* (New York: Thames and Hudson, 1987), 388–401; Gilda de Mello e Souza, *O espírito das roupas* (São Paulo: Companhia das Letras, 1987); Duarte, "Pássaros e cientistas," 5–11.

15. Heitor Pereira da Cunha, *Viagens e caçadas em Mato Grosso* (Rio de Janeiro: Francisco Alves, 1922); Alberto de Carvalho, *Manual do Caçador* (São Paulo: privately printed, 1924), 26–28; Bento Arruda, *Por campos e mattas* (São Paulo: Editora Monteiro Lobato, 1925); Bernardo de Castro, *Tiro ao voo* (Rio de Janeiro: privately printed, 1925). On the hunting of small game by the poor as a

supplementary food source in the early twentieth century, see Arthur Neiva, "Prefácio," in Eurico Santos, *Pássaros do Brasil*, 8–9, 2nd ed. (Rio de Janeiro: F. Briguiet, 1948); Mário Olivério Pinto, "Resultados ornithológicos de uma excursão pelo oeste de São Paulo e sul de Mato Grosso," *Revista do Museu Paulista* 17 (1932): 689–826; Ihering, "Proteção às aves," 316.

16. José Bento Monteiro Lobato, *Caçadas de Pedrinho*, Série Literatura Infantil 9 (São Paulo: Companhia Editora Nacional, 1933), 7–15. It should be pointed out that Monteiro Lobato's thoughts on the question of nature—just as his thoughts on cruelty toward animals—were quite complex, as evinced in his 1921 short story "*Homo Sapiens*." Lobato, *A onda verde* (São Paulo: Globo, 2008), 70–75. Monteiro Lobato's (1882–1948) writings continue to stir passions and fuel controversy on topics like race, politics, and energy resources even today. On Monteiro Lobato, see Carmen Lucia de Azevedo, Marcia Camargos, and Vladimir Saccheta, *Furacão na Botocúndia* (São Paulo: SENAC, 1997).

17. Verena Alberti, "O século do moderno: Modos de vida e consumo na República," in *A República no Brasil*, ed. Angela de Castro Gomes, Dulce Pandolfi, and Verena Alberti, 263–64, 285 (Rio de Janeiro: Nova Fronteira, Editora FGV, 2002); Bóris Fausto, *História do Brasil*, 12th ed. (São Paulo: Edusp, 2004), 292.

18. Ernst Mayr, *What Makes Biology Unique? Considerations on the Autonomy of a Scientific Discipline* (Cambridge: Cambridge University Press, 2004), 3–4. The two other scholars, in addition to Lamarck, were the Germans Gottfried Reinhold Treviranus and Karl Friedrich Burdach.

19. Michel Foucault, *The Order of Things: An Archaeology of the Human Sciences* (New York: Vintage, 1994), 137–38. See also Foucault, "Cuvier's Position in the History of Biology," *Critique of Anthropology* 4:13–14 (1979): 125–30. For his critique of the history of science, see Foucault, *The Archaeology of Knowledge* (New York: Pantheon, 1982), 135–40. On Cuvier, Foucault, and biology, see Regina Horta Duarte, "Limites e fronteiras entre história e biologia em Michel Foucault: As palavras e as coisas e o surgimento da biologia no século XIX," in *Cartografias de Foucault*, ed. Durval Albuquerque, Alfredo Veiga-Netto, and Alipio Souza Filho, 343–54 (Belo Horizonte: Autêntica, 2008). On Cuvier's role in the emergence of contemporary biology, see Ernst Mayr, *The Growth of Biological Thought: Diversity, Evolution and Inheritance*, 11th ed. (Cambridge, MA: Harvard University Press, 2000), 363–70; Stephen Jay Gould, *The Structure of Evolutionary Theory*, 15th ed. (Cambridge, MA: Harvard University Press, 2002), 295.

20. Foucault, *Security, Territory, Population*, 77–79. The biologist Ernst Mayr argues that one of the most important concepts introduced by Darwin was "population thinking," or "biopopulation." Emphasizing differences between individuals in the same species, this notion made a break with typological thinking, or constant types, which tied ideas about living organisms to explanations of the physical world: "This view was a totally new philosophical concept, crucial for the understanding of the theory of natural selection" (Mayr, *What Makes Biology Unique?*, 88).
21. Lynn Nyhart, "Natural History and the 'New' Biology," in *Cultures of Natural History*, ed. N. Jardine, J. A. Secord, and E. C. Spary, 441 (Cambridge: Cambridge University Press, 1996). See also Joseph Caron, "Biology in the Life Sciences: A Historiographical Contribution," *History of Science* 26 (1988): 223–68; Peter J. Bowler, *The Earth Encompassed* (New York: Norton, 1993), 248–58, 306–35.
22. Roquette-Pinto, "Euclydes da Cunha, naturalista," *Revista do Brasil* 8:29 (1918): 20.
23. In 1982, Mayr stated that "only in our age is it being appreciated what a great conceptual contribution natural history has made to biology" (*The Growth of Biological Thought*, 143). It is interesting to note that this revered ornithologist and evolutionist, and outstanding figure of the so-called evolutionary synthesis, felt compelled to defend the notion that biology is just as true a science as physics or chemistry. See Mayr, *What Makes Biology Unique?*, 11–20.
24. On the relation between natural history and the construction of the great seventeenth- and eighteenth-century empires, see Richard Drayton, *Nature's Government: Science, Imperial Britain, and the "Improvement" of the World* (New Haven: Yale University Press, 2000); Richard Grove, *Green Imperialism: Colonial Expansion, Tropical Island Edens and the Origins of Environmentalism, 1600–1860* (New York: Cambridge University Press, 1997).
25. Philip Pauly, *Biologists and the Promise of American Life* (Princeton: Princeton University Press, 2002), 3–7, 10, 166–69. John Dewey was a lodestar for a number of Brazilian educational thinkers in the 1920s and 1930s, especially the educator Anísio Teixeira. The collection Atualidades Pedagógicas, edited by Fernando Azevedo as a series within the Biblioteca Pedagógica (discussed in chap. 2), released the titles *Como pensamos* (1933; How we think) and *Democracia e educação* (1936; Democracy and education). On Anísio Teixeira (1900–1971) as a leader of the movement for educational renewal in Brazil, see Carlos Monarcha, *Anísio Teixeira: Obra de uma vida* (Rio de Janeiro: DP&A, 2001);

Diana Gonçalves Vidal, *O exercício disciplinado do olhar: Livros, leituras e práticas de formação docente no Instituto de Educação do Distrito Federal* (Bragança Paulista: Editora USF, 2001).

26. Pauly, *Biologists and the Promise of American Life*, 9, 171–89; Gregg Mitman, *The State of Nature: Ecology, Community and American Social Thought, 1900–1950* (Chicago: University of Chicago Press, 1992), 7, 45–71.

27. The Escola Nova (new school) was a Brazilian movement for educational renewal that was heavily influenced by the U.S. thinker John Dewey and by the progressive education movement. It defended free secular public education and scientific and experimental teaching for young people. On the Escola Nova movement in the 1930s, see Manoel Lourenço Filho, *Introducção ao estudo da Escola Nova* (São Paulo: Melhoramentos, 1930); Fernando Azevedo et al. *Manifesto dos Pioneiros da Educação Nova: A reconstrução educacional no Brasil ao povo e ao governo* (São Paulo: Companhia Editora Nacional, 1932). Anísio Teixeira engaged in a clear dialogue with Dewey in other works, such as *Educação progressiva* (1933; Progressive education), published as part of the Atualidades Pedagógicas collection, which also featured Portuguese translations of Dewey. During his persecution under the Estado Novo regime, Teixeira was forced to stop publishing. His book *Educação para a democracia* (Education for democracy) was only released in 1953; it came out as volume 57 of the Atualidades Pedagógicas collection and inaugurated a series of other publications of his works. See Maria Rita de Almeida Toledo, "Coleção 'Atualidades Pedagógicas': Do projeto político ao projeto editorial (1931–1981)" (PhD diss., Universidade de São Paulo, 2001), 82–100.

28. Stuart McCook, *States of Nature: Science, Agriculture and Environment in the Spanish Caribbean, 1760–1940* (Austin: University of Texas Press, 2002), 5, 20–28, 128–29.

29. McCook, *States of Nature*, 22; McCook, "Crônica de uma praga anunciada: Epidemias agrícolas e história ambiental do café nas Américas," *Varia Historia* 24:39 (2008): 87–112; John Soluri, *Banana Cultures: Agriculture, Consumption, and Environmental Change in Honduras and the United States* (Austin: University of Texas Press, 2005), 104–27. On the blight that assailed the rubber-tapping regions of the Amazon in the twentieth century, see Warren Dean, *Brazil and the Struggle for Rubber* (Cambridge: Cambridge University Press, 2002), 53–66.

30. Roquette-Pinto, *Ensaios de anthropologia brasiliana*, Coleção Brasiliana 22 (São Paulo: Companhia Editora Nacional, 1933), 5, 69.

31. Francis Galton, *Hereditary Genius: An Inquiry Into its Laws and Consequences* (London: Macmillan and Co., 1869), iii-iv; Galton, *Natural Inheritance* (London: Macmillan and Co., 1889), 1–3; Mayr, *The Growth of Biological Thought*, 47, 695, 784–85; Gould, *The Structure of Evolutionary Theory*, 344–46.
32. On the New World polemic—the idea that plants and animals in the New World were inferior to those in the Old—see the classic work by Antonello Gerbi, *The Dispute of the New World: The History of a Polemic* (Pittsburgh: University of Pittsburgh Press, 2010), 3–34. See also Monteiro Lobato's interpretation, published in 1918, of how this controversy was settled in Brazil through the public health movement: *Mr. Slang e o Brasil, Problema Vital*, 12th ed. (São Paulo: Brasiliense, 1968), 321–28.
33. Mello Leitão, "Pela eugenia," *O Imparcial*, Rio de Janeiro, 1922, newspaper clipping, BR.MN.JF.O.MN.DR.2, SEMEAR. On views of eugenics in Brazilian intellectual circles during these years, see Tania de Luca, *A Revista do Brasil: Um diagnóstico para a (N)ação* (São Paulo: Editora Unesp, 1999), 131–201, 223–30; Nísia Trindade Lima, *Um sertão chamado Brasil* (Rio de Janeiro: Revan, 1999), 115–17; Lilia Schwarcz, *O espetáculo das raças: Cientistas, instituições e questão racial no Brasil, 1870–1930* (São Paulo: Companhia das Letras, 1993), 215–17, 235–38; João Italo de Oliveira e Silva, "Por uma eugenia latino-americana: Victor Delfino e Renato Kehl" (MA thesis, Universidade Federal de Minas Gerais, 2008). On how physicians fought the use of wet nurses because they were worried about so-called racial regeneration, see Okezi T. Otovo, "From Mãe Preta to Mãe Desamparada, Maternity and Public Health in Post-Abolition Bahia," *Luso-Brazilian Review* 48:2 (2011): 164–91. There is a vast bibliography on the postabolition period in Brazil and the rise of racism and the exclusion of people of color. See especially Sidney Chalhoub, *Trabalho, lar e botequim* (Campinas: Editora da Unicamp, 2001); Hebe Mattos and Ana Lugão Rios, *Memórias do cativeiro: Família, trabalho e cidadania no pós-abolição* (São Paulo: Civilização Brasileira, 2005).
34. José Bento Monteiro Lobato, *Urupês*, 30th ed. (São Paulo: Brasiliense, 1984), 139–55; Lobato, *Mr. Slang e o Brasil*, 298. On the impact of the character Jeca Tatu on race debates in Brazil and on Monteiro Lobato's change in argument, see Luca, *A Revista do Brasil*, 202–15.
35. José Bento Monteiro Lobato, *Problema vital, Jeca Tatu e outros textos* (São Paulo: Editora Globo, 2010), 102–11.
36. Jaime Benchimol, *Manguinhos, do sonho à vida: A ciência na Belle Époque* (Rio de Janeiro: Editora Fiocruz. 1990), 5–88; Jaime Benchimol and Luiz Antonio

Teixeira, *Cobras, lagartos e outros bichos: Uma história comparada dos institutos Oswaldo Cruz e Butantan* (Rio de Janeiro: Editora Fiocruz, 1993), 13–109; Lucia Bulcão, Almir El-Kareh, and Jane Sayd, "Ciência e ensino médico no Brasil," *História, Ciências, Saúde—Manguinhos* 14:2 (2007): 470. On the Oswaldo Cruz Institute and its scientists, see Stepan, *Beginnings of Brazilian Science*, 65–104.

37. Benchimol, *Manguinhos, do sonho à vida*, 14; Jaime Benchimol, "Domingos José Freire e os primórdios da bacteriologia no Brasil," *História, Ciências, Saúde—Manguinhos* 2:1 (1995): 67–98. See also the entry "Instituto Bacteriológico Domingos Freire," *Dicionário Histórico-Biográfico das Ciências da Saúde no Brasil, 1832–1930* (Rio de Janeiro: Fundação Oswaldo Cruz), http://www.dichistoriasaude.coc.fiocruz.br, accessed June 29, 2009. On medicine in Brazil, see Heloisa Starling, Ligia Germano, and Rita de Cássia Marques eds., *Medicina, história em exame* (Belo Horizonte: Editora UFMG, 2011).

38. Schwartzman, *Um espaço para a ciência*, 108–10. Adolpho Lutz (1855–1940), Brazilian, studied medicine in Switzerland and specialized at the Pasteur Institute in Paris. He was "the most experienced and versatile member of a small group of physicians" who pioneered tropical medicine and zoology applied to medicine in Brazil, especially in the field of epidemiology and the study of infectious disease. Jaime Benchimol, "Adolpho Lutz: Um esboço biográfico," *História, Ciências, Saúde—Manguinhos* 10:1 (2003): 13–83. See also Jaime Benchimol and Magali Romero Sá, eds., *Adolpho Lutz, obra completa* (Rio de Janeiro: Editora Fiocruz, 2006).

39. Benchimol and Teixeira, *Cobras, lagartos e outros bichos*, 13–17; Nara Britto, *Oswaldo Cruz: A construção de um mito na ciência brasileira* (Rio de Janeiro: Editora Fiocruz, 1995).

40. Marta de Almeida, "São Paulo na virada do século XX: Um laboratório de saúde pública para o Brasil," *Tempo* 10:19 (2005): 77–89.

41. N. Lima, *Um sertão chamado Brasil*, 79–89. On Arthur Neiva (1880–1943) and the respect he earned among his contemporaries, see Lobato, *Mr. Slang e o Brasil*, 169–76. On Belisário Penna (1868–1939), see Eduardo Thielen and Ricardo Santos, "Belisário Penna: Notas fotobiográficas," *História, Ciências, Saúde—Manguinhos* 9:2 (2002): 387–404.

42. Lobato, *Mr. Slang e o Brasil*, 247–48, 257.

43. Foucault, *Society Must Be Defended*, 243–44. As Foucault saw it, what happened was "state control of the biological" (239).

44. Sidney Chalhoub, *Cidade febril: Cortiços e epidemias na Corte Imperial* (São Paulo: Companhia das Letras, 1996), 50.
45. Roquette-Pinto, *Ensaios de anthropologia*, 1–16, 105.
46. Nelson Sanjad, "Da 'abominável profissão de vampiros': Emílio Goeldi e os mosquitos no Pará (1905)," *História, Ciências, Saúde—Manguinhos* 10:1 (2003): 85–111; Benchimol, "Adolpho Lutz," 14, 58, 60, 65.
47. Schwarcz, *O espetáculo das raças*, 84–89; Fausto, *História do Brasil*, 291–94.
48. Lopes, *O Brasil descobre a pesquisa*, 248–64; Mello Leitão, *A biologia no Brasil*, Coleção Brasiliana 99 (São Paulo: Companhia Editora Nacional, 1937), 189–90. On the relation between natural and sociopolitical aspects of the rise and fall of rubber in the early twentieth century in Brazil, see Dean, *Brazil and the Struggle for Rubber*, 53–66.
49. The Higher School of Agricultural Science and Veterinary Medicine first opened its doors in the city of Rio in 1913. In 1934, it became the National School of Agronomy, and since 1963 it has been part of the Federal Rural University of Rio de Janeiro (Universidade Federal Rural do Rio de Janeiro, or UFRRJ). André Felipe Cândido da Silva, "A campanha contra a broca-do-café em São Paulo (1924–1927)," *História, Ciências, Saúde—Manguinhos* 13:4 (2006): 957–93; "Escola Superior de Agricultura e Medicina Veterinária," in *Dicionário Histórico-Biográfico das Ciências da Saúde no Brasil* (Rio de Janeiro: Casa de Oswaldo Cruz), www.dichistoriasaude.coc.fiocruz.br, accessed June 23, 2009. Angelo Moreira da Costa Lima (1887–1964), a native of the city of Rio, also graduated from medical school; he worked at the Oswaldo Cruz Institute and later at the National School of Agronomy. One of Brazil's greatest entomologists and an expert in the field of agricultural entomology, he published the twelve-volume *Insetos do Brasil*. See Pedro Bloch, *Vida e obra de Angelo Moreira da Costa Lima*, Coleção Vultos da Ciência Brasileira 1 (Rio de Janeiro: Conselho Nacional de Pesquisas, 1968); Marcio Ferreira Rangel, "Um entomólogo chamado Costa Lima: A construção de um saber e a consolidação de um patrimônio científico" (PhD diss., Fundação Oswaldo Cruz, Rio de Janeiro, 2006); Mario Aragão, "Angelo Moreira da Costa Lima: De candidato a cirurgião a entomologista maior," *Cadernos de Saúde Pública* 4:4 (1988): 353–55.
50. On the controversial figure of Edmundo Navarro de Andrade (1881–1941), see Augusto Martini, "O plantador de eucaliptos: A questão da preservação florestal no Brasil e o resgate documental do legado de Edmundo Navarro de Andrade" (MA thesis, Universidade de São Paulo, 2004); Warren Dean, *With*

Broadax and Firebrand: Destruction of the Brazilian Atlantic Forest (Berkeley: University of California Press, 1997), 236–38, 245–46. For a very favorable appraisal of Navarro, see also Lobato, *Onda verde*, 53–57.

51. A. Silva, "A campanha contra a broca," 981–89; Márcia Rebouças, "Pelo resgate da memória documental das ciências e da agricultura: O acervo do Instituto Biológico de São Paulo," *História, Ciências, Saúde—Manguinhos* 13:4 (2006): 995–1005.

52. *O Museu Nacional: Commemoração do primeiro centenário da Independência do Brasil, notas e informações* (Rio de Janeiro: Imprensa Nacional, 1922), 37–47.

53. Lopes, *O Brasil descobre a pesquisa*, 93–127, 225–47; Mello Leitão, *A biologia no Brasil*, 186–87.

54. Dominichi Miranda de Sá, Magali Romero Sá, and Nísia Trindade Lima, "Telégrafos e inventário do território no Brasil: As atividades científicas da Comissão Rondon (1907–1915)," *História, Ciências, Saúde—Manguinhos* 15:3 (2008): 779–810.

55. On Alípio de Miranda Ribeiro (1874–1939), a self-taught zoologist from Minas Gerais and an expert on vertebrates, see José Pombal Junior, "Nota biográfica sobre Alípio de Miranda Ribeiro," *Revista Brasileira de Zoologia* 19:3 (2002): 935–39. On Hoehne (1882–1959), who was a member of the expeditions of 1908, 1910, 1912, and 1913, see José Luiz de Andrade Franco and José Augusto Drummond, "Frederico Carlos Hoehne: A atualidade de um pioneiro no campo da proteção à natureza no Brasil," *Ambiente e Sociedade* 8:1 (2006): 141–66.

56. Roquette-Pinto, *Rondônia*, 4th ed., Coleção Brasiliana 39 (São Paulo: Companhia Editora Nacional, 1935), 88–90. On the fascinating figure of Cândido Rondon (1856–1958), see Todd Diacon, *Stringing Together a Nation: Cândido Mariano da Silva Rondon and the Construction of a Modern Brazil, 1906–1930* (Durham: Duke University Press, 2004). Diacon offers a thought-provoking critique of the revisionist analysis of Rondon set out by Antônio Lima, *Um grande cerco de paz* (Petropolis: Vozes, 1995).

57. Mello Leitão, *A biologia no Brasil*, 258. Freire Alemão (1797–1874) contributed invaluable research to Brazilian botany by collecting and describing species during major scientific expeditions, like the Scientific Commission of the Empire (Comissão Científica do Império) of 1859–61. See Rita de Cássia de Jesus Morais, "Nos verdes campos da ciência: A trajetória acadêmica do médico e botanista brasileiro Francisco de Freire Alemão" (MA thesis, Fundação Oswaldo Cruz, Rio de Janeiro, 2005); Lorelai Kury, *Comissão Científica do império* (Rio de Janeiro: Andrea J. Editorial, 2009). Lauro Travassos (1890–1970) was

a graduate of the Rio de Janeiro School of Medicine and worked at the Oswaldo Cruz Institute. His greatest contribution was to the field of zoology, with papers of international note on helminthology and entomology. See Luiz Fernando Ferreira, "Lauro Travassos," *Cadernos de Saúde Pública* 5:4 (1989): 461–69. On the researcher Carlos Chagas (1879–1934), another graduate of the Rio de Janeiro School of Medicine, see Simone Kropf, *Doença de Chagas, doença do Brasil: Ciência, saúde e nação 1909–1962* (Rio de Janeiro: Editora Fiocruz, 2009).

58. Ihering, *Proteção às aves*, 326–329.
59. Mello Leitão, *A biologia no Brasil*, 190. On the fascinating trajectory of Emília Snethlage, a German ornithologist who came to Brazil in 1905 and never left, along with a reference to the food episode, see Miriam Junghans, "Emília Snethlage (1868–1929): Uma naturalista alemã na Amazônia," *História, Ciências, Saúde—Manguinhos* 15, suppl. (2008): 243–55; Roquette-Pinto, *Ensaios brasilianos*, Coleção Brasiliana 190 (São Paulo: Companhia Editora Nacional, 1941), 88–90.
60. Silva, "A campanha contra a broca," 990.
61. Benchimol, *Cobras, lagartos e outros bichos*, 164–66.
62. Benchimol, *Manguinhos, do sonho à vida*, 60–65.
63. Pandolfi, "Os anos 1930," 18–20; Fausto, *História do Brasil*, 333.
64. Mello Leitão, "Protegendo a nossa fauna"; Sampaio, *Phytogeographia do Brasil*, Coleção Brasiliana 35 (São Paulo: Companhia Editora Nacional, 1934), 12–17.
65. Maria Helena Capelato, "O Estado Novo: O que trouxe de novo?" in *O Brasil Republicano: O tempo do nacional-estatismo*, ed. Lucília Delgado and Jorge Ferreira, 107–43 (Rio de Janeiro: Civilização Brasileira, 2003), 112.
66. Pandolfi, "Os anos 1930," 23–30; Angela de Castro Gomes, "Confronto e compromisso no processo de constitucionalização, 1930–1935," in *O Brasil Republicano*, ed. Bóris Fausto, 8–75, 2nd ed., Coleção História Geral da Civilização Brasileira 10 (São Paulo: Difel, 1983).
67. Roquette-Pinto, "Ofício 391." In 1932, the museum had forty-one employees all told, janitor included.
68. Mello Leitão, "O estudo da história natural 1–2," *O Imparcial*, April 11–12, 1923; Mello Leitão, "Doutorandos de 1908," *Jornal do Commércio*, December 5, 1948; Mello Leitão, "Vida e obra científica de Alberto José de Sampaio e Afrânio Peixoto," *Jornal do Commércio*, December 16, 1947. Albert Dastre (1844–1917) was an important French physiologist. On João Joaquim Pizarro, who was widely rebuffed by his contemporaries for his radical Darwinism, see Regina

Gualtieri, "O evolucionismo na produção científica do Museu Nacional do Rio de Janeiro (1876–1915)," in *A recepção do darwinismo no Brasil*, ed. Heloisa Maria Bertol Domingues, Magali Romero Sá, and Thomas Glick, 65–66 (Rio de Janeiro: Editora Fiocruz, 2003).

69. Mello Leitão, "O livro de minha vida para ser lido pelos meus netos," MS, 1946–47, 61 pp., 11–13, Cândido de Mello Leitão papers, Academia Brasileira de Ciências (hereafter ABC).

70. Mello Leitão, "O livro de minha vida," 18–20; Edison Correa and Sebastião Gusmão, *85 anos da Faculdade de Medicina da UFMG* (Belo Horizonte: Coopmed, 1996), 3–4. See also José Luiz de Andrade Franco and José Augusto Drummond, "Cândido de Mello Leitão: As ciências biológicas e a valorização da natureza e da diversidade da vida," *História, Ciências, Saúde—Manguinhos* 14:4 (2006): 1265–90; Adriano Kury and Renner Baptista, "Arachnological Papers Published by Cândido Firmino de Mello-Leitão (Arachnida)," *Publicações Avulsas do Museu Nacional* 105 (Rio de Janeiro: Museu Nacional, 2004).

71. On the Rio de Janeiro Normal School in the 1920s and 1930s, see Liéte Accácio, "A Escola Normal do Rio de Janeiro," *Revista de Pedagogia* 2:5 (2009): 4; Heloisa Rocha, "Prescrevendo regras de bem viver: Cultura escolar e racionalidade científica," *Cadernos Cedes* 20:52 (2000): 55–73.

72. Mello Leitão, "O estudo da história natural 1–2"; Mello Leitão, "A methodologia da história natural," *O Estado de S. Paulo*, July 1, 1927. For a thorough study of the Brazilian Education Association (ABE), see Marta Maria Chagas de Carvalho, *Molde nacional e fôrma cívica* (Bragança Paulista: Editora USF, 1998). Mello Leitão was president of the ABE in 1924–25, 1929–30, and 1932–34 and a member of the board of directors in 1926–32 and 1935–36.

73. Mello Leitão, "Vida e obra científica dos acadêmicos," 4–5.

74. Mello Leitão, *Compêndio de botânica* (Rio de Janeiro: Francisco Alves, 1924); Osório Duque Estrada, "Compêndio de botânica," *Jornal do Brasil*, March 28, 1925.

75. Roquette-Pinto, *Ensaios brasilianos*, 104–95.

76. Sampaio held the position of first secretary of the ABC during the 1933–35 term and of vice president in 1939–41; Mello Leitão served as vice president in 1937–39 and president in 1943–45.

77. Ildeu Moreira, Luisa Massarani, and Jayme Aranha, "Roquette-Pinto e a divulgação científica," in *Antropologia brasiliana: Ciência e educação na obra de Edgard Roquette-Pinto*, ed. Nísia Trindade Lima and Dominichi Miranda de Sá, 247–70 (Rio de Janeiro: Editora Fiocruz; Belo Horizonte: Editora UFMG

2008); Vera Roquette-Pinto, "Roquette-Pinto, o rádio e o cinema educativos," *Revista USP* 56 (2002–3): 10–15.

78. Capanema, *A natureza no projeto de construção*, 20–44.
79. Roquette-Pinto, *Rondônia*, 104, 199.
80. Lima, Sá, and Sá, *Telégrafos e inventário*, 794; Capanema, *A natureza no projeto de construção*, 37; Mário Ferri, "A botânica no Brasil," in *As ciências no Brasil*, ed. Fernando Azevedo, 2:149–200 (São Paulo: Melhoramentos, 1956).
81. Rachel Crotman, "A hora do cinema, entrevista com Roquette-Pinto," *Diário de Notícias*, August 27, 1933.
82. Nísia Trindade Lima and Dominichi Miranda de Sá, "Roquette-Pinto e sua geração na república das letras e da ciência," in *Antropologia brasiliana*, ed. Lima and Sá, 57–63; Vanderlei Sebastião de Sousa, *Edgard Roquette-Pinto e o retrato antropológico brasileiro, 1905–1935* (Natal: Editora UFRN; Rio de Janeiro: Editora FGV, 2015), 310–11.
83. Capanema, *A natureza no projeto de construção*, 37; José Luiz de Andrade Franco and José Augusto Drummond, "Alberto José Sampaio: Um botânico brasileiro e o seu programa de proteção à natureza," *Varia Historia* 21:33 (2005): 135; Mello Leitão, *Vida e obra científica*, 4–5.
84. Mello Leitão, "O livro de minha vida," 24; "A viagem de estudos de um pediatra brasileiro: Parte amanhã para a Europa o Dr. Mello Leitão," *A Vanguarda*, April 12, 1926.
85. Roquette-Pinto, "Depoimento," March 23, 1939, as quoted in Lima and Sá, "Roquette-Pinto e sua geração," 59; "Os cursos na Associação Brasileira de Educação: O professor Roquette-Pinto encerrou as suas conferências sobre anthropologia," *Electron* 1:14 (1926): 4.
86. On Alberto Torres (1865–1917), see Adalberto Marson, *A ideologia nacionalista em Alberto Torres* (São Paulo: Duas Cidades, 1979). Torres's main works are *A organização nacional* (Rio de Janeiro: Imprensa Nacional, 1914) and *O problema nacional brasileiro* (Rio de Janeiro: Imprensa Nacional, 1914).
87. Roquette-Pinto, "Alberto Torres," *Revista Nacional de Educação* 2:18–19 (1934): 1–6; Roquette-Pinto, "Prefácio," in Mello Leitão, *A biologia no Brasil*, 10. Note that in this particular book, Mello Leitão included chapters on a number of areas within biology, like botany, zoology, physiology, anatomy, and anthropology.
88. Lima and Sá, *Antropologia brasiliana*, 14.
89. This accounts for a good number of Roquette-Pinto's contradictory stances, according to Giralda Seyferth, "Roquette-Pinto e o debate sobre raça e imigração no Brasil," in *Antropologia brasiliana*, ed. Lima and Sá, 170–72.

90. Diana Gonçalves Vidal, "Manifiesto de los pioneros de la educación nueva," *Transatlántica de Educación* 5 (2008): 50–63. See also Monarcha, *Brasil Arcaico, Escola Nova*.
91. Foucault, *Society Must be Defended*, 243–44; Azevedo et al. *Manifesto dos Pioneiros da Educação Nova*.
92. These activities are analyzed in chapters 2 and 3.
93. Kury and Baptista, "Arachnological Papers," 3. Some works were published posthumously.
94. Getúlio Vargas, "Educar," *Revista Nacional de Educação* 1:11–12 (1933): 1–9.
95. Francisco Campos, "A Revista Nacional de Educação: Carta manuscrita reproduzida," *Revista Nacional de Educação* 1:1 (1933): 3.
96. Angela de Castro Gomes and Martha Abreu, "A nova 'Velha' República," *Tempo* 13:26 (2009): 3.
97. Campos, "A Revista Nacional de Educação," 3.
98. Helena Bomeny, "Novos talentos, vícios antigos: Os renovadores e a política educacional," *Estudos Históricos* 6:11 (1993): 24–39. On Francisco Campos and his actions in the field of education, see Ana Maria Casassanta Peixoto, "A escola no projeto de construção do Brasil moderno: A reforma Francisco Campos em Minas Gerais," *Educação em Revista* 16:12 (1992): 12–17; Maria Célia Moraes, "Educação e política nos anos 30: A presença de Francisco Campos," *Revista Brasileira de Estudos Pedagógicos* 73:174 (1992): 201–321.
99. Washington Pires, "Discurso de posse na pasta da Educação e Saúde Pública," *Revista Nacional de Educação* 1:1 (1932): 1–2.
100. Alcir Lenharo, *A sacralização da política* (São Paulo: Papirus, 1986), 16, 35.
101. Roquette-Pinto, "O Brasil e a anthropogeographia," *Revista do Brasil* 3:12 (1916): 323; Mello Leitão, "Dostoievsky e a revolução russa," *O Imparcial*, June 22, 1922; Mello Leitão, "Ensino universitário na Rússia," *O Imparcial*, March 17, 1922. Colmeia (beehive) was a student society to which the brothers Carlos and Edgard Süssekind de Mendonça belonged; they would later work with Roquette-Pinto at Rádio Sociedade. Pedro Gouvea Filho, "E. Roquette-Pinto, antropólogo e educador," *Revista Brasileira de Estudos Pedagógicos* 24:59 (1955): 35.
102. Lenharo, *A sacralização da política*, 22–23. On the fading of the question of class struggle from historical memory and much of historiography in the period around 1930, see De Decca, *1930: O silêncio dos vencidos*, 75–101; Marilena Chauí, "História a contrapelo," in De Decca, *1930: O silêncio dos vencidos*, 11–30.
103. In the inaugural class of the course in physiology that Roquette-Pinto taught

at the National University of Asuncion (Universidad Nacional de Asunción) in 1920, he defined the organism as "a harmonious, indivisible whole," made up of parts that are "functionally integrated and correlated" and joined together for one sole purpose. Roquette-Pinto, *Conceito actual da vida* (Rio de Janeiro: Süssekind de Mendonça, 1922), 39, 47.

104. Sampaio, *Phytogeographia do Brasil*, 249.

105. The Escola de Recife (school of Recife) was a cultural and intellectual movement of graduates of what was then the Recife School of Law (Faculdade de Direito do Recife; currently part of the Federal University of Pernambuco). The group was founded as a forum for debating new Western philosophical ideas in order to define Brazil's road to modernization. See Angela Alonso, *Ideias em movimento: A geração de 1870 na crise do Brasil Império* (São Paulo: Paz e Terra, 2002), 133–42.

106. Heloisa Maria Bertol Domingues, "Controvérsias evolucionistas no Brasil do século XIX," in *A recepção do darwinismo no Brasil*, 97–123; Gualtieri, "O evolucionismo na produção científica," 45–96.

107. Mello Leitão, *A vida maravilhosa dos animais*, Série Iniciação Científica 7 (São Paulo: Companhia Editora Nacional, 1935), 197–227.

108. The same tendency could be felt in the broader context of European and U.S. scientific circles in the early twentieth century. In the 1920s, some scientists went back to Darwin's theory of natural selection and reconciled it with Mendelian genetics. The so-called evolutionary synthesis began in the 1920s but gained its strongest momentum in the wake of works published in 1936–47 that established neo-Darwinism, according to which evolution should take a number of factors into account: mutation, recombination, natural selection, migration, and genetic drift. Gould, *The Structure of Evolutionary Theory*, 503; Mayr, *The Growth of Biological Thought*, 566–73; Michael Barbour, "Ecological Fragmentation in the Fifties," in *Uncommon Ground: Rethinking the Human Place in Nature*, ed. William Cronon, 233–55 (New York: W. W. Norton, 1996).

109. Mello Leitão, "Genética," *Revista Nacional de Educação* 2:16–17 (1934): 30–35.

110. Mello Leitão, "Teia da vida," *O Imparcial*, October 27, 1922; Mello Leitão, "O mais velho dos poetas . . . ," *O Imparcial*, newspaper clipping, BR.MN.JF.O.MN.DR.2, SEMEAR; Mello Leitão, "Cinco minutos de conversación con la más alta autoridad em arañas, entrevista," *La Libertad*, newspaper clipping, BR.MN.JF.O.MN.DR.3, SEMEAR. Mello Leitão's view of evolution would change over the years. In the early 1920s, he focused more on genetic determinism, whereas in the 1930s, his discourse had more in common with

advocates of the potential of adaptation, which, in the human case, he saw as enabled by education and hygiene. I believe this was partially due to changes in Brazilian intellectual circles but likewise to his close ties to Roquette-Pinto. When it comes to these men's interpretations of Darwin, it is always wise to remember that—according to *On the Origin of Species*—evolution is a noncontinuous, indeterminate, nonlinear, nonprogressive, and nonteleological process, as pointed out by both Steven J. Gould, *Ever Since Darwin: Reflections on Natural History* (New York: W.W. Norton, 1977), 56–62, and Mayr, *What Makes Biology Unique?*, 39–66.

111. Edgard Roquette-Pinto, "Euclydes da Cunha, naturalista," 26, 27; Roquette-Pinto, "O Brasil e a anthropogeographia," 324; Roquette-Pinto, *Ensaios de anthropologia brasiliana*, 59, 85–87, 94–95. It is worth noting that the reduction of Darwin's theories to the struggle for existence and survival of the fittest was then prevalent among these scientists. Yet other complex, thought-provoking readings and interpretations of Darwin's thought are possible, as biologists like Mayr and Gould have shown. See also Regina Horta Duarte, "História e biologia: Diálogos possíveis, distâncias necessárias," *História, Ciências, Saúde—Manguinhos* 16:4 (2009): 927–40; Duarte, "Limites e fronteiras entre história e biologia," 343–54. These debates resonated among Brazil's intellectuals and yielded major new lines of interpretation about the country. Gilberto Freyre was an attentive reader of Roquette-Pinto's neo-Lamarckist anthropology, as pointed out by Maria Lúcia Pallares-Burke, *Gilberto Freyre: Um vitoriano dos trópicos* (São Paulo: Editora Unesp, 2005), 332–44.

112. See Decree 23.672, of January 2, 1934. The remaining members of the council would be representatives of the following bodies and groups: the Game and Fish Service (Serviço de Caça e Pesca), fishers and hunters, owners of fishing vessels, fish and seafood canneries, and the navy. Four of the members were also supposed to be recognized experts in their field.

113. Teresa Urban, *Saudade do matão: Relembrando a história da conservação da natureza no Brasil* (Curitiba: Editora UFPR, Fundação O Boticário, Fundação MacArthur, 1998), 4; Drummond, *A legislação ambiental brasileira*, 132–35; Osny Duarte Pereira, *Direito florestal brasileiro* (Rio de Janeiro: Borsoi, 1950), 131–32.

114. The Law on Scientific Expeditions was enacted through Decree 22.698, of January 26, 1934; the Water Code, through Decree 24.643, of July 10, 1934; and the law stipulating animal protection measures, through Decree 24.645, of July 10, 1934. Alberto José de Sampaio, ed., "Primeira Conferência Brasileira de

Proteção à Natureza," *Boletim do Museu Nacional* 11:2 (1935): 61–68. Sampaio was the first National Museum scientist appointed to the post.

115. On the start and end dates for ministers' terms of office under Vargas, see Fábio Koifman, ed., *Presidentes do Brasil* (Rio de Janeiro: Cultura, Editora Rio, 2002), 330–63.

116. Mello Leitão, *A biologia no Brasil*, 140, 164–65, 177, 187–88. The second minister of education and public health was Washington Pires; Mello Leitão probably slipped up here, meaning to refer to the third minister, Gustavo Capanema, who held the post from July 1934 to October 1945. On Capanema, see Schwartzman, Bomeny, and Costa, *Tempos de Capanema*.

117. Mello Leitão, *Zoogeografia do Brasil*, Coleção Brasiliana 77 (São Paulo: Companhia Editora Nacional, 1937). The museum's appointment of a staff member to the post can be found in Roquette-Pinto to Ministro da Agricultura Odilon Duarte Braga, September 14, 1934, MN.MN.DR, SEMEAR. See also Mello Leitão, "O livro de minha vida," 28–29.

CHAPTER 2

1. Edgard Roquette-Pinto, "Seção de Assistência ao Ensino da História Natural do Museu Nacional: Relatório de 1930," SAE 146–5, SEMEAR. The dual-purpose conference and exhibit hall was inaugurated in 1927.

2. Roquette-Pinto, "Relatório dos trabalhos do Museu Nacional em 1930," BR.MN.MN.DR.1931, SEMEAR. The Ministry of Education and Public Health (MESP) was created on November 14, 1930, and the museum was attached to it in March 1931.

3. Roquette-Pinto, "Relatório dos trabalhos do Museu Nacional em 1931," BR.MN.BR.MN.DR.1932, SEMEAR. On seeing and statecraft, see James C. Scott, *Seeing Like a State: How Certain Schemes to Improve the Human Condition Have Failed* (New Haven: Yale University Press, 1998), 1–8. On "the optic of the State" as a new way of seeing and its relation to "national being," see also Andermann, *The Optic of the State*, 24.

4. Roquette-Pinto, "Apresentação," *Uiára* 1:1 (1937): 3.

5. Paulo Roquette-Pinto, "O Museu Nacional," *Uiára* 1:1 (1937): 6.

6. Edgard Roquette-Pinto, "Radio Educação no Brasil," *Electron* 1:6 (1926): 15; Roquette-Pinto, *Seixos rolados* (Rio de Janeiro: Mendonça Machado, 1927), 234–35.

7. Roquette-Pinto, as quoted in Moreira, Massarani, and Aranha, "Roquette-Pinto e a divulgação científica," 248. On the "collective soul," see Roquette-Pinto, "O Brasil e a anthropogeographia."
8. My approach draws from Foucault, for whom power is always relational and characterized not only by repression or exclusion but especially by the creation of realities and practices. In the specific case of science, Foucault also stresses that it is not about "knowing what external power imposes itself on science" as much as of knowing "what effects of power circulate among scientific statements." Michel Foucault, *Discipline and Punish: The Birth of the Prison* (New York: Vintage, 1995), 194; Foucault, "Truth and Power," in *Power*, vol. 3 of *Essential Works of Foucault, 1954–1984*, ed. James D. Faubion, 111–33 (New York: New Press, 2001).
9. Angela de Castro Gomes, *A República, a história e o IHGB* (Belo Horizonte, Brasil: Argvmentvm, 2009), 95–104.
10. "Alto Falante," *Electron* 1:1 (1926): 2. For online access to the documents of Rádio Sociedade, see the website of the Projeto Memória da Rádio Sociedade coordinated by Luísa Massarani and Nísia Trindade Lima, http://www.fiocruz.br/radiosociedade/cgi/cgilua.exe/sys/start.htm?sid=35, accessed December 18, 2009. On Roquette-Pinto and radio, see Renato de Souza Gilioli, "Educação e cultura no rádio brasileiro: Concepções de radioescola em Roquette-Pinto" (PhD diss., Universidade de São Paulo, 2008).
11. The Brazilian Academy of Sciences was founded in 1916 under the name Brazilian Society of Sciences (Sociedade Brasileira de Ciências); the new name was chosen in 1921. "The history of the founding of the society is not very clear when it comes to stipulating who its founders were. Although the act that established the organization was signed by only fifteen people, and twenty-five scientists were part of the body's initial core, some references mention thirty-eight participants. Most were doctors or engineers: Adalberto Menezes de Oliveira, Alberto Betim Paes Leme, Alberto Childe, Alfredo Lisboa, Alix Corrêa de Lemos, Allypio de Miranda Ribeiro, Allyrio Hugueney de Mattos, André Gustavo Paulo de Frontin, Ângelo Moreira da Costa Lima, Antônio Ennes de Souza, Arthur Alexandre Moses, Bruno Alvares da Silva Lobo, Carlos Ernesto Julius Lohmann, Daniel Henninger, Domingos Fernandes da Costa, Edgard Roquette-Pinto, Euzébio Paulo de Oliveira, Everardo Adolpho Backheuser, Francisco Xavier de Oliveira Menezes, Guilherme Florence, Henrique Beaurepaire Rohan Aragão, Henrique Charles Morize, Ignacio Manoel Azevedo do Amaral, João Alberto Constantino Löffgren, Joaquim

Cândido da Costa Senna, John Casper Branner, Juliano Moreira, Júlio Cezar Diogo, Lélio Itapuambyra Gama, Licínio Athanásio Cardoso, Luiz de Carvalho e Mello, Luiz Gonzaga de Campos, Manoel Bonfim, Manuel Amoroso Costa, Mario Rodrigues de Souza, Oswaldo Gonçalves Cruz, Sebastião Sodré da Gama, and Theophilus Henry Lee. "Sociedade Brasileira de Ciências," *Dicionário histórico-biográfico das ciências da saúde no Brasil (1832–1930)*.

12. Ildeu Moreira and Luisa Massarani, "A divulgação científica no Rio de Janeiro: Algumas reflexões sobre a década de 20," *História, Ciências, Saúde—Manguinhos* 7:3 (2000): 327–51; Dominichi Miranda de Sá, *A ciência como profissão: Médicos, bacharéis e cientistas no Brasil* (Rio de Janeiro: Editora Fiocruz, 2006), 164–65, 181; José Jerônimo de Alencar Alves, "As ciências na Academia e as expectativas de progresso e modernização, Brasil 1916–1929," in *Espaços da ciência no Brasil 1800–1930*, ed. Maria Amélia Dantes, 190–96 (Rio de Janeiro: Editora Fiocruz, 2001).

13. Roquette-Pinto, *Seixos rolados*, 232. Every issue of *Electron* featured articles on experimental matters and radio. See, e.g., Henrique Morize, "Aparelho automático para a recepção do sinal S.O.S.," *Radio* 23 (1934): 9; "A polarização horizontal das ondas curtas," *Electron* 1:9 (1926): 15; "O mais simples e mais econômico aparelho receptor de galena," *Electron* 1:10 (1926): 10; "Como construir receptores," *Electron* 1:18 (1926): 6; Theodore Naken, "Fotografia e reprodução do som," *Electron* 1:16 (1926): 7–8.

14. The engineer and astronomer Henrique Morize (1860–1930) was born in France but in 1874 immigrated to Brazil, where he eventually became a Brazilian citizen. For part of his career, he taught at the Polytechnic School, an engineering institute founded in Rio de Janeiro in 1792.

15. When broader political turmoil led to the declaration of a state of siege in July 1922, radio broadcasting was banned. Roquette-Pinto and his colleagues from the academy were among those who fought the measure. To judge from the advertisements placed by importers and retailers of loudspeakers, tubes, receivers, radio sets, and other equipment featured in the magazines *Radio* and *Electron*, major business interests felt trampled; this was likely the biggest deciding factor when the ban was modified in 1923, making it possible to found Rádio Sociedade. Roquette-Pinto, *Ensaios brasilianos*, 66–75; V. Roquette-Pinto, "Roquette-Pinto, o rádio e o cinema educativos"; Gouvea Filho, "E. Roquette-Pinto"; "A maior estação de radiofonia da América do Sul," *Radio* 1:16 (1924): 40; Gilioli, *Educação e cultura no rádio*, 138–42.

16. This publishing experience was preceded by initiatives undertaken by other

members of the board of Rádio Sociedade, Carlos Süssekind de Mendonça and Francisco Venâncio, who owned Livraria Científica, a bookstore that had started out as a meeting place for intellectuals, including Roquette-Pinto. See Gilioli, *Educação e cultura no rádio*, 99, 197. In 1922, the bookstore moved into publishing with the release of the collection Cultura Contemporânea (Contemporary culture), devoted to "the ideas of time and the knowledge most indispensable to the human spirit." The first volume was written by Roquette-Pinto, while subsequent volumes were by Amoroso Costa, Miguel Osório de Almeida, Henrique Morize, Juliano Moreira (all from the ABC), Afranio Peixoto, Alceu Amoroso Lima, and Moraes Coutinho. See Roquette-Pinto, *Conceito actual da vida*, back cover.

17. "O terceiro aniversário da Rádio Sociedade do Rio de Janeiro," *Electron* 1:6 (1926): 4.
18. "Notícias," *Radio* 1:16 (1924): 2; "Modelos para obter licença de installação de um apparelho receptor radio-telephonico," *Radio* 1:16 (1926): 3; Moreira and Massarani, "A divulgação científica."
19. "Alto falante," *Electron* 1:4 (1926): 1.
20. Roquette-Pinto, *Seixos rolados*, 236–37; Roquette-Pinto, "O Brasil e a radiocultura," *Radio* 1:19 (1924): 9; "Alto Falante," *Electron* 1:2 (1926): 2. On the parallel between these men's optimism about new technologies and today's excitement about the Internet, and on their real limits as panaceas, see Moreira, Massarani, and Aranha, *Roquette-Pinto e a divulgação científica*, 267.
21. Roquette-Pinto, *Seixos rolados*, 239–40.
22. Roquette-Pinto, *Seixos rolados*, 238–39.
23. Gilioli, *Educação e cultura no rádio*, 204–372. On the donation of Rádio Sociedade to the government and the desire to protect it from the Press and Propaganda Department by attaching it to the MESP, see esp. 185–86, 327–29.
24. Gilioli, *Educação e cultura no rádio*, 292, 274–75.
25. "Estação de rádio escola no Distrito Federal," *Revista Nacional de Educação* 2:16–17 (1934): 44–51. Manuel Lourenço Filho (1897–1970) put through a public education reform in the state of Ceará in 1922–23 and held top posts with the MESP in the early 1930s. Along with Anísio Teixeira, Fernando Azevedo, and Roquette-Pinto, he was one of the signers of the Escola Nova's Manifesto of the Pioneers of New Education.
26. Sampaio's forestry course was divided into the following modules: general ideas on forests and their importance; how to properly plant a tree; guidelines on arboriculture in Brazil; how commercial forests are raised; how native

forests renew themselves; land for reforestation; best trees for planting; cultivating and operating commercial forests; mixed and homogeneous forests; and pruning, harvesting, and gross and net income from commercial forests. "Os cursos da Rádio Sociedade," *Electron* 1:1 (1926): 7. See also Capanema, "A natureza no projeto de construção," 141–42.

27. It is interesting to note that Mello Leitão offered his listening and reading audiences (his lectures later came out in print) quite a favorable view of Wegener's continental drift theory, formulated in 1915 but utterly rejected by scientists at that time. It was only in the 1950s, after the use of submarines had led to new research findings, that geologists revived and embraced the theory. Mello Leitão, "A gênese dos continentes e oceanos segundo Wegener," *Revista Nacional de Educação* 2:15 (1933): 49–54.
28. Ivan Domingues, ed., *Conhecimento e transdisciplinaridade* (Belo Horizonte: Editora UFMG, 2005), 25–26.
29. Lopes, *O Brasil descobre a pesquisa*, 14, 53, 70. On the National Museum in the nineteenth century, see also José Neves Bittencourt, "Território largo e profundo: Os acervos dos museus do Rio de Janeiro como representação do Estado Imperial 1808–1889" (PhD diss., Universidade Federal Fluminense, Niterói, 1997), 89–220.
30. Drayton, *Nature's Government*, xv.
31. Lopes, *O Brasil descobre a pesquisa*, 21–25, 325; Heloisa Maria Bertol Domingues and Magali Romero Sá, "O museu nacional e o ensino das ciências naturais no Brasil no século XIX," *Revista da Sociedade Brasileira para o Progresso da Ciência* 15 (1996): 82.
32. George B. Goode, as quoted in Sally G. Kohlstedt, "Thoughts in Things: Modernity, History, and North American Museums," *Isis* 96:4 (2005): 588. On some of the earliest signs of this U.S. movement in favor of the experimental study of nature—where research institutes, museums, universities, and schools were all involved and the work of Dewey found great response—see chap. 1, n. 25, above, and Sally G. Kohlstedt, "Nature, Not Books: Scientists and the Origins of the Nature-Study Movement in the 1890s," *Isis* 96:3 (2005): 324–52; Kohlstedt, "A Better Crop of Boys and Girls: The School Gardening Movement, 1890–1920," *History of Education Quarterly* 48:1 (2008): 58–93. On Roquette-Pinto and the American Museum of Natural History, see Gilioli, *Educação e cultura no rádio*, 116–18.
33. Roquette-Pinto, "Relatório dos trabalhos do Museu Nacional em 1929," BR.MN.MN.DR.1930, SEMEAR.

34. Roquette-Pinto, *Ensaios brasilianos*, 156–58.
35. Roquette-Pinto, "Serviço de Assistência ao Ensino da História Natural, relatório de 1929, folhas 7–12," SAE 146.5, Doc. 7.09A, folder 105, SEMEAR.
36. Roquette-Pinto, "Serviço de Assistência ao Ensino da História Natural, relatório de 1929, folha 14." On the broad diversity of courses offered, see "Relatórios do Serviço de Assistência ao Museu entre 1929 e 1937," SAE 146.5, SEMEAR. See also Sá, *A ciência como profissão*, 178; Gilioli, *Educação e cultura no rádio*, 117.
37. "Reação salutar," *O Paiz*, November 9, 1929.
38. Roquette-Pinto to Fernando Magalhães, Reitor da Universidade do Distrito Federal, February 3, 1932, BR.MN.MN.DR.1932, SEMEAR. The anthropologist Heloísa Alberto Torres (1895–1977) was one of the first women on the staff of the National Museum (along with Bertha Lutz); she directed the institution from 1938 to 1955. See Heloisa Domingues, "Heloísa Alberto Torres e o inquérito nacional sobre ciências naturais e antropológicas, 1946," *Boletim do Museu Paraense Emílio Goeldi, Ciências Humanas* 5:3 (2010): 625–43.
39. Sampaio, *Biogeographia dynamica: A natureza e o homem no Brasil*, Coleção Brasiliana 53 (Rio de Janeiro: Companhia Editora Nacional, 1935), 9, 13, 47, 66, 203. On Bertha Lutz (1894–1976), daughter of Adolpho Lutz, see Lia Sousa, Mariana Sombrio, and Margareth Lopes, "Para ler Bertha Lutz," *Cadernos Pagu* 24 (2005): 315–25.
40. Mello Leitão, "Papel educativo do Museu Nacional de História Natural," *Revista Nacional de Educação* 1:2 (1932): 96–98.
41. On the idea that life in America is inferior, as preached by France's Comte de Buffon (1707–88), and on Thomas Jefferson's reaction, see Lee A. Dugatkin, *Mr. Jefferson and the Giant Moose: Natural History in Early America* (Chicago: University of Chicago Press, 2009). On the various moments in the New World polemic, see Gerbi, *The Dispute of the New World*.
42. Roquette-Pinto, *Seixos rolados*, 31–43.
43. On the idea of Brazil as a showy spectacle, I draw from the analysis of historical museums by Ulpiano Meneses, "A exposição museológica e o conhecimento histórico," in *Museus: Dos gabinetes de curiosidades à museologia moderna*, ed. Betânia Gonçalves Figueiredo and Diana Gonçalves Vidal, 15–84 (Belo Horizonte: Argvmentvm, 2005). On time and exhibits, see Martha Marandino, "Museus de Ciências como espaços de educação," in *Museus*, ed. Figueiredo and Vidal, 166. On science museums and cognitive experience, see Yael Bamberger and Tali Tal, "The Learning Environment of Natural History Museums:

Multiple Ways to Capture Students' Views," *Learning Environments Research* 12 (2009): 115–29; Christian Heath and Dirk Vom Lehn, "Configuring 'Interactivity': Enhancing Engagement in Science Centres and Museums," *Social Studies of Science* 38:1 (2008): 63–91. On museums and the shaping of collective identities, see José Mauro Loureiro, "Museu de ciência, divulgação científica e hegemonia," *Ciência da Informação* 32:1 (2003): 88–95. On science museum exhibits as a means of communication, see Marandino, "Museus de ciências," 166; Maria Esther Valente, Sibele Cazelli, and Fátima Alves, "Museus, ciência e educação: Novos desafios," *História, Ciência, Saúde – Manguinhos* 12, suppl. (2005): 183–203.

44. Roquette-Pinto, "O Brasil e a anthopogeographia," 331.
45. "Um cinema de filmes educativos no Museu Nacional," *Cinearte* 8:317 (March 23, 1923); Crotman, "A hora do cinema." See also Sheila Schvarzman, *Humberto Mauro e as imagens do Brasil* (São Paulo: Editora UNESP, 2004), 115.
46. On the potential offered by the historical analysis of objects at science museums and the networks of relations among collectors, scientists, and visitors that grow out of collecting, organizing, and exhibiting these objects, see Samuel Alberti, "Objects and the Museum," *Isis* 96:4 (2005): 559–72.
47. These same tendencies can also be detected in the early twentieth-century republican practices developed during the publication of the *Almanaque Garnier* and the construction of a "pedagogy of nationality." See Dutra, *Rebeldes literários da República*, 232.
48. Roquette-Pinto, "Comentários," *Revista Nacional de Educação* 1:3 (1932): 1; Mário Augusto Teixeira de Freitas, "Primeira ronda," *Revista Nacional de Educação* 2:11–12 (1933): 63–64.
49. The observations that follow are taken from a modified version of Regina Horta Duarte, "Em todos os lares, o conforto moral da ciência e da arte: A Revista Nacional de Educação e a divulgação científica no Brasil (1932–1934)," *História, Ciências, Saúde—Manguinhos* 11:1 (2004): 33–56.
50. Jonathas Serrano, "Carta a Roquette-Pinto," *Revista Nacional de Educação* 1:8 (1933): 2–4. See also Schwarzman, *Humberto Mauro*, 128; Anísio Teixeira, "Instruções regulando a censura e seleções de filmes educativos," *Revista Nacional de Educação* 1:9 (1933): 19–29. For some examples of contemporary criticisms about the alleged lenience of the Commission for the Censorship and Selection of Educational Films, which approved the release of movies containing scenes some considered immoral, see "Censura cinematographica," *Diário de Notícias*, June 8, 1933; "Existe uma comissão de censura," *Avante*, March 8, 1934.

51. German naturalists Carl Friedrich Philipp von Martius (1794–1868) and Johann Baptist von Spix (1781–1826) were in Brazil from 1917 to 1920. The first Portuguese translation of the full text of their *Reise in Brasilien* came out in 1938 (*Viagem ao Brasil*). Karen Lisboa, *A nova Atlântida de Spix e Martius: Natureza e civilização na viagem pelo Brasil* (São Paulo: Hucitec, 1997). On Alexandre Rodrigues Ferreira (1756–1815) and his expedition through the Brazilian Amazon in 1783–92, under order to Queen D. Maria I of Portugal, see Maria Elice Prestes, *A investigação da natureza no Brasil colônia* (São Paulo: Annablume, 2000), 80–93.
52. Cynthia Greive Veiga, "Educação estética para o povo," in *500 anos de educação no Brasil*, ed. Eliana Marta et al., 399–422 (Belo Horizonte: Autêntica, 2000).
53. Roquette-Pinto, "O cinema e a educação popular no Brasil," *Revista Nacional de Educação* 1:5 (1933): 1–9. One conto de réis was equal to 1 million réis.
54. Lima, *Um sertão chamado Brasil*, 60; Alvaro Fonseca, "Discurso inaugural na Sociedade Carioca de Educação," *Revista Nacional de Educação* 1:4 (1933): 1–3; Afranio Peixoto, "Discurso pronunciado no banquete oferecido ao prof. Miguel Pereira, em 19 de maio de 1918," in *Afrânio versus Afrânio*, ed. P. Leão et al., 29–37 (Niterói: Tipografia "Jeronimo Silva," 1922).
55. Mario Augusto Teixeira de Freitas, "Estatística e educação," *Revista Nacional de Educação* 1:1 (1933): 56–59. As a pioneer of statistics in Brazil, Teixeira de Freitas (1890–1956) focused especially on education in Brazil and published reference books, such as *O ensino primário no Brasil* (São Paulo: Melhoramentos, 1934) and *O que dizem os números sobre ensino primário* (São Paulo: Melhoramentos, 1937).
56. Sampaio, "O babassu," *Revista Nacional de Educação* 1:6 (1933): 37–43; Sampaio, "A carnaúba," *Revista Nacional de Educação* 1:7 (1933): 54–59.
57. One example of these articles is the publication of excerpts by the French entomologist Eugène Louis Bouvier: "Formigas agricultoras," *Revista Nacional de Educação* 1:4 (1933): 36–45.
58. Regina Horta Duarte, "Entre fábulas y arcanos: Proyecciones políticas y sociales del pensamiento biológico sobre la naturaleza en Brasil," in *Naturaleza en declive: Miradas a la historia ambiental de América Latina y el Caribe*, ed. Reinaldo Funes, 317–46 (Valencia: Centro Francisco Tomás y Valiente UNED; Alzira-Valencia: Fundación de Historia Social, 2008).
59. Otto Frensel was editor of *Boletim do Leite*, a magazine that played a vital role in publicizing the "science of nutrition" in Brazil in the 1930s through the work of physicians who were proponents of eugenics. Sören Brinkmann, "Milk and

Modernity: Ideology and Nutrition Policies during the Vargas Era," *História, Ciências, Saúde—Manguinhos* 21:1 (2014): 263–80. On meteorologist Sampaio Ferraz, see Solange Godoy, *O avô do tempo: Diário de um meterologista* (Rio de Janeiro: EMC, 2009).

60. "O céu do Brasil," *Revista Nacional de Educação* 1:4 (1933): 80–84. On the Cruls Mission, see Moema Vergara, "Ciência e história no Relatório da Comissão Exploradora do Planalto Central na Primeira República," *História, Ciências, Saúde—Manguinhos* 13:4 (2006): 909–25.
61. Othelo Reis and Jonathas Serrano both taught at Pedro II high school—geography and history, respectively. Serrano published books on the teaching of history, including *Methodologia da História na aula primária* (Rio de Janeiro: Francisco Alves, 1917), and *Como se ensina História* (São Paulo: Melhoramentos, 1935).
62. Alberto Childe, "A leitura dos hieróglifos," *Revista Nacional de Educação* 1:2 (1932): 24. On Childe, see Luis de Castro Faria, *Alberto Childe: O homem, a obra e a sua época*, Publicações Avulsas do Museu Nacional 60 (Rio de Janeiro: Museu Nacional, 1970).
63. Raimundo Lopes, "Antropogeografia," *Revista Nacional de Educação* 1:11–12 (1933): 17–23. See also Luiz de Castro Faria, "Um sábio maranhense no Museu Nacional," in *Raimundo Lopes, dois estudos resgatados*, ed. Heloisa Maria Bertol Domingues and Alfredo Almeida, 7–13 (Rio de Janeiro: Ouro sobre Azul, 2010).
64. Roquette-Pinto, *Seixos rolados*, 216.
65. Fernando Guerra Duval, "Palestras sobre fotografia," *Revista Nacional de Educação* 1:4 (1933): 64–69. Guerra Duval was editor in chief of the magazine *Photorama*, one of Brazil's first publications on amateur photography.
66. Pedro Sinzig, "Como ouvir música," *Revista Nacional de Educação* 1:4 (1933): 11–14. Friar Sinzig (1876–1956) wrote in the press and was active in book publishing in Brazil, having founded a number of magazines and newspapers. He also wrote books about music, including *Cancioneiro de modinhas populares* (Rio de Janeiro: B. Herder, 1914), and *O Brasil cantando* (Petropolis: Vozes, 1938).
67. "Primeiras letras," *Revista Nacional de Educação* 1:2 (1932): 80.
68. "Comentário," *Revista Nacional de Educação* 1:10 (1933): 21; Maria G. R. Almeida, "Utilinda brincando ou o ensino da leitura e da escrita ao alcance do leigo e da própria criança pela sonografia," *Revista Nacional de Educação* 1:10 (1933): 22–32.

69. Roquette-Pinto, *Ensaios brasilianos*, 63.
70. Some of the titles of Roquette-Pinto's works derive from this proposal, like *Ensaios brasilianos* and *Ensaios de antropologia brasiliana*.
71. Eliana Dutra, "A nação nos livros: A biblioteca ideal na coleção Brasiliana," in *Política, Nação e Edição*, ed. Eliana R. de Freitas Dutra and Jean Yves Mollier, 299–314 (São Paulo: AnnaBlume, 2006).
72. The original Portuguese titles of the collections in the Biblioteca Pedagógica were *Iniciação Científica, Livros Didáticos, Literatura Infantil, Atualidades Pedagógicas*, and *Brasiliana*.
73. Heloisa Pontes, "Retratos do Brasil: Editores, editoras e 'Coleções Brasiliana' nas décadas de 30, 40 e 50," in *História das Ciências Sociais no Brasil*, ed. Sergio Miceli, 359–409 (São Paulo: Vértice, Revista dos Tribunais, 1989); Maria Rita Toledo, "O projeto político-cultural da Coleção Atualidades Pedagógicas," in *Política, Nação e Edição*, ed. Dutra and Mollier, 335–50.
74. Marta Maria Chagas de Carvalho and Maria Rita de Almeira Toledo, "Reforma escolar, pedagogia da Escola Nova e usos do impresso," *Contemporaneidade e Educação* 5:7 (2000): 71–91.
75. Sampaio, *Phytogeographia do Brasil*, Coleção Brasiliana 35 (São Paulo: Companhia Editora Nacional, 1934). See also José Luiz de Andrade Franco and José Drummond, "Wilderness and the Brazilian Mind I: Nation and Nature in Brazil from the 1920s to the 1940s," *Environmental History* 13 (2008): 724–50.
76. Roquette-Pinto, "Prefácio," in Sampaio, *Biogeographia dynamica*, 5–6.
77. Sampaio, "Primeira Conferência Brasileira de Proteção à Natureza," *Boletim do Museu Nacional* 11:1 (1935): 1–230. On the conference, see also Capanema, *A natureza no projeto de construção*, 27; Franco and Drummond, *Proteção à natureza*, 43–64; José Luiz de Andrade Franco and José Augusto Drummond, "Wilderness and the Brazilian Mind II: The First Brazilian Conference on Natural Protection," *Environmental History* 14 (2009): 82–102.
78. Sampaio, ed., *Programa da Primeira Conferência Brasileira de Proteção à Natureza* (Rio de Janeiro: Sociedade dos Amigos das Árvores, 1933), 1–7.
79. Roquette-Pinto, "Prefácio," in Sampaio, *Biogeographia dynamica*, 5–6. Brazilian social thought had long been concerned with the protection of nature; major authors had devoted themselves to the topic over the course of the nineteenth century, as shown by José Augusto Pádua, *Um sopro de destruição: Pensamento político e crítica ambiental no Brasil escravista, 1786–1888* (Rio de Janeiro: Jorge Zahar, 2004). Authors from the National Museum, however, were engaged less with nineteenth-century thought than with their contemporaries, like

Euclides da Cunha and Alberto Torres, perhaps because they were eager to make a clean break with Brazil's imperial past.

80. Sampaio, ed., "Primeira Conferência Brasileira de Proteção à Natureza," *Boletim do Museu Nacional* 11:2 (1935): 48–49, 115.
81. Sampaio, *Biogeographia dynamica*, 8.
82. Sampaio, "Primeira Conferência Brasileira," 48.
83. Sampaio, *A alimentação sertaneja e do interior da Amazônia: Onomástica da alimentação rural*, Coleção Brasiliana 238 (São Paulo: Companhia Editora Nacional, 1944), 7–27.
84. Sampaio, *A alimentação sertaneja*, 329–41. This book covers a plethora of topics and proposals and is certainly deserving of special study. To give an example that is still relevant today, on p. 95 Sampaio suggests that popular restaurants be opened to offer well-balanced menus at affordable prices. On Pereira Barreto, physician and positivist, see the interesting text by Angela Alonso, "O positivismo de Luís Pereira Barreto e o pensamento brasileiro no final do século XIX" (São Paulo: Instituto de Estudos Avançados da Universidade de São Paulo, 1995), http://www.iea.usp.br/publicacoes/textos/alonsopositivismo.pdf, accessed October 17, 2009.
85. Afonso Taunay, "Prefácio," in Mello Leitão, *Visitantes do Primeiro Império*, Coleção Brasiliana 32 (São Paulo: Companhia Editora Nacional, 1934), 9–11. Mello Leitão used accounts by Luiz de Freycinet, Maria Graham, Louis Isodore Duperrey, Barão de Bougainville, Laplace, and Charles Darwin.
86. Mello Leitão, *Zoogeografia do Brasil*, 7–8.
87. Roquette-Pinto, "Prefácio," in Mello Leitão, *A biologia no Brasil*, 9–12.
88. The intention of creating a reference work is also apparent in 1946, when the same editor published Mello Leitão's 646-page biology dictionary, the first of its kind in Portuguese. Mello Leitão, *Glossário biológico: Pequeno dicionário de termos técnicos empregados em ciências biológicas, botânica, ecologia, genética, zoologia* (São Paulo: Companhia Editora Nacional, 1946).
89. Gaspar de Carvajal, Alonso Rojas, and Cristobal Acuña, *Descobrimentos do Rio das Amazonas*, trans. and ed. Cândido de Mello Leitão, Coleção Brasiliana 203 (São Paulo: Companhia Editora Nacional, 1941).
90. Mello Leitão, "Prefácio," in Henry Bates, *O naturalista no rio Amazonas*, trans. Cândido de Mello Leitão, Coleção Brasiliana 237–37A (São Paulo: Companhia Editora Nacional, 1944), 5.
91. Mello Leitão, *Curso elementar de história natural*, 4 vols. Série Livros Didáticos 3, 15, 17 and 36 (São Paulo: Companhia Editora Nacional, 1933–35); Mello

Leitão, *Biologia geral*, Série Livros Didáticos 42 (São Paulo: Companhia Editora Nacional, 1940).

92. Múcio Leão, "Registro literário: A vida maravilhosa dos animais," *Jornal do Brasil*, clipping news, BR.MN.JF.O.MN.DR.3, SEMEAR; Lucia Miguel Pereira, "Livros: A vida maravilhosa dos animais," *Gazeta de Notícias*, August 25, 1945; Maurício de Lacerda, "O comedor de escorpiões, comensalismo, parasitismo, imperialismo," *O Imparcial*, August 9, 1935. These thoughts on *A vida maravilhosa dos animais* first appeared in Regina Horta Duarte, "Biologia, natureza e República no Brasil nos escritos de Mello Leitão (1922–1945)," *Revista Brasileira de História* 29:58 (2009): 317–40.
93. Colin Favret, "Jean-Henri Fabre: His Life Experiences and Predisposition against Darwinism," *American Entomologist* 45:1 (1999): 38–48; Patrick Tort, *Fabre, le miroir aux insectes* (Paris: Éditions Vuibert, 2002).
94. Mello Leitão, *A vida maravilhosa dos animais*, 37, 63.
95. Mello Leitão, *A vida maravilhosa dos animais*, 27, 155.
96. Mello Leitão, *La vida em la selva*, Biblioteca de Autores Brasileños Traducidos al Castellano 12 (Buenos Aires: Imprenta López, 1949).
97. Foucault, *The Archaeology of Knowledge*, 23.
98. Roquette-Pinto, "Rádio Educação do Brasil," 15.
99. The complete text of Roquette-Pinto's letter to Mário de Andrade, dated September 8, 1934, can be found in Schvarzman, *Humberto Mauro e as imagens do Brasil*, 130. See also Elisangela Galvão, "A ciência vai ao cinema: Uma análise de filmes educativos e de divulgação científica do Instituto Nacional do Cinema Educativo (INCE)" (MA thesis, Universidade Federal do Rio de Janeiro, 2004, 49). The letters of praise from the magazine's readers can be found in the Roquette-Pinto papers, Arquivo do Museu Nacional. On the significance of this shift in censorship duties from the MESP to the Ministry of Justice and Internal Affairs, see Marialva Barbosa, *História Cultural da Imprensa* (Rio de Janeiro: Mauad, 2007), 117–25.
100. Alberto Betim Paes Leme, "Relatório dos Trabalhos do Museu Nacional referentes ao anno de 1936, dirigido ao Exmo. Snr. Dr. Gustavo Capanema," January 11, 1937, 1, BR.MN.MN.DR, SEMEAR.
101. Mello Leitão, *A biologia no Brasil*, 140, 164–65, 177, 188. On the original plans of Teixeira and Roquette-Pinto for PRD-5, see Gilioli, "Educação e cultura no rádio," 274–92.
102. Capanema, "A natureza no projeto de construção"; Sampaio, "Ofício ao

Ministerio da Educação," BR.MN.AJS.DB.10, SEMEAR. I would like to thank Carolina Capanema for calling my attention to this document.

103. Similarly, despite their productive friendship, Anísio Teixeira and Fernando Azevedo held diverse views on education, as shown by Vidal, *O exercício disciplinado do olhar*, 79–84.

104. Schwartzman, Bomeny, and Costa, *Tempos de Capanema*, 19.

105. Jackson de Figueiredo (1891–1928) and Alceu Amoroso Lima (1893–1983) were influential Catholic thinkers in Brazil. Figueiredo died young, but Amoroso Lima carried on his legacy; in addition to heading the magazine *A Ordem* and the Dom Vital Center, he worked to gain a bigger voice for Catholics in the public sphere in the 1930s. During elections to the Constituent Assembly, he founded the Catholic Voter League (Liga Eleitoral Católica), which supported all candidates who promised to defend church stances, such as religious instruction in schools. Cândido Moreira Rodrigues, *A Ordem: Uma revista de intelectuais católicos* (São Paulo: Fapesp, 2005); Fernando Antonio Pinheiro Filho, "A invenção da ordem: Intelectuais católicos no Brasil," *Tempo Social* 19:1 (2007): 33–49.

106. On the two main groups that made up the Brazilian Education Association from its outset and on the 1932 rupture, see Carvalho, *Molde nacional e fôrma cívica*, 43, 71, 112–13.

107. Fausto, *Historia do Brasil*, 340–52; Pandolfi, "Os anos 1930, as incertezas do regime," 13–38; Iglesias, *Trajetória política do Brasil*, 231–46.

108. Schwartzman, Bomeny, and Costa, *Tempos de Capanema*, 61–64, 71–78.

109. Independent of the question of compliance with the nationwide standards put in place by Capanema, the University of São Paulo had a very specific project that the minister dared not challenge. The establishment of the university was how São Paulo's powerful agricultural and industrial elites had responded to their state's defeat during the Constitutionalist Revolution of 1932, an uprising against the Vargas administration. The new university valued teaching alongside scientific and technological research; trained the state's intellectual and scientific elites; and grounded itself in autonomous, secular instruction. From its inception, the University of São Paulo recruited renowned researchers, like Claude Lévi-Strauss, Roger Bastide, and Fernando Braudel, among others. Schwartzman, *Um espaço para a ciência*, 164–79; Marshall Eakin and Paulo Almeida, *Envisioning Brazil: A Guide to Brazilian Studies in the United States* (Madison: University of Wisconsin Press, 2005), 381–89.

110. Schwartzman, Bomeny, and Costa, *Tempos de Capanema*, 221–29. It should be pointed out that the broader political culture under Minister Capanema was complex and multifaceted, as shown by Williams, *Culture Wars in Brazil*, 14–15, 81.
111. Schwartzman, Bomeny, and Costa, *Tempos de Capanema*, 194–95.

CHAPTER 3

1. "Nota," *Boletim do Museu de Biologia Prof. Mello Leitão*, Mello Leitão papers, ABC. Augusto Ruschi (1915–86) was one of Brazil's greatest hummingbird experts, as well as a conservation activist, but little research has been done on him. See André Ruschi, "Augusto Ruschi: Feitos e legados," www.augustoruschi.com.br, accessed May 5, 2015; Mileide Formigni and Hilton Silva, "Conservação ambiental e populações tradicionais: Uma contribuição bibliográfica da obra de A. Ruschi," *Boletim do Museu de Biologia Mello Leitão* 31 (2013): 50–75. Mello Leitão identified Silvestri as being from the "Reggio Laboratorio di Entomologia Agraria di Portici," but I could find no record of any such laboratory; he was more likely with the Portici agricultural college's Laboratorio di Zoologia Generale e Agraria. For a link to some papers by Silvestri published in the laboratory's journal, see http://www.biodiversitylibrary.org/itemdetails/48991, accessed July 7, 2015.
2. "Sociedade dos Amigos do Museu Nacional," *Jornal do Commércio*, July 25, 1937. On the Guinle family, see the entry "Guilherme Guinle," www.cpdoc.fgv.br, accessed October 12, 2009; Geraldo Mendes Barros, *Guilherme Guinle, ensaio biográfico* (Rio de Janeiro: Agir, 1982); Jacques Marcovitch, *Pioneiros e empreendedores: A saga do desenvolvimento no Brasil* (São Paulo: Edusp, 2005), 155–92.
3. *Estatutos: Sociedade dos Amigos do Museu Nacional* (Rio de Janeiro: M.E.S. Serviço Gráfico, 1937), MN.500.1074.S.678, SEMEAR. A handwritten copy of the draft bylaws, drawn up by Mello Leitão, is found in the same archive folder; it shows that the bylaws were approved almost in full, with only a few minor alterations.
4. Roquette-Pinto, *Seixos rolados*, 101. Uiára is a mythological creature of Tupi origin, incorporated into Brazilian folklore in the form of a siren with a magical song.
5. Roquette-Pinto, "Apresentação."

6. Augusto Ruschi to Sociedade dos Amigos do Museu Nacional, December 6, 1939, MN.500.1074-S.678, SEMEAR; Secretaria da Agricultura do Espírito Santo to Heloísa Alberto Torres, September 2, 1940, MN.500.1074.S.678, SEMEAR.
7. The Society of the Friends of the National Museum was refounded in 1959 by the arachnologist José Lacerda de Araujo Feio (1912–73), a disciple of Mello Leitão. Its journal is still published today.
8. "A situação no Museu Nacional," *A Noite*, 1946, clipping news, BR.MN.JF. O.MN.DR.2, SEMEAR. See also *O Museu Nacional* (São Paulo: Banco Safra, 2007), 30–41. The National Museum became part of the University of Brazil in 1946, and its exhibits were reopened to the public the following year. See also Thereza Zavaro, *Os diretores do Museu Nacional/UFRJ* (Rio de Janeiro: Museu Nacional, 2008).
9. Sampaio, *A alimentação sertaneja*; Jorge Marcgrave, *História Natural do Brasil*, trans. Alberto J. de Sampaio (São Paulo: Imprensa Oficial do Estado, 1942). For biographical information on Sampaio, see Capanema, *A natureza no projeto de construção*, 30–33.
10. Mello Leitão, "Vida e obra científica," 4–5.
11. Schvarzman, *Humberto Mauro e as imagens*, 132.
12. Luciano de Feo was director of L'Unione Cinematografica Educativa, in Italy, from 1925 to 1928. He participated in the creation of the first Venice Film Festival in 1932 (then called the Esposizione d'Arte Cinematografica), and he was the first director of the festival. See Fiamma Lussana, "Cinema 'Educatore': Luciano de Feo direttore dell'Istituto Luce," *Studi Storici* 56:4 (2015): 935–62.
13. The original titles of the films were *Vacina contra a raiva*, *Céu do Brasil*, and *Victoria Regia* (1937); *Vacina contra febre amarela* and *O puraquê* (1939); and *Coração físico de Ostwald* (1942). The titles of the films made in conjunction with Humberto Mauro were *O descobrimento do Brasil* (1937), *Bandeirantes* (1940), and *Argila* (1942). To view some of the shorts made at the INCE from 1936 to 1966, see http://www.bcc.org.br/filmes/ince, accessed July 22, 2015. For an analysis of the films made by the INCE and relations between Roquette-Pinto and Humberto Mauro, see Schvarzman, *Humberto Mauro e as imagens*, 137–244; Galvão, "A ciência vai ao cinema," 97–145.
14. The profession of biologist was only made official quite recently in Brazil, regulated under Law 6.684, September 3, 1979.
15. Federico Daus, "Prólogo," in Mello Leitão, *La vida en la selva*, 7–12.

16. His first two publications were Mello Leitão, "Alguns gêneros e espécies novas de araneidos do Brasil," *Brotéria, Série Zoologia* 13:2–3 (1915): 129–42; and "Notas arachnologicas, notas à margem do catálogo-índice de Petrunkevitch," *Brotéria, Série Zoologia* 13:2–3 (1915): 143–44.
17. When Mello Leitão passed away at the age of sixty-two, one of the messages of condolences received by the ABC came from arachnologist Lucien Berland, with the National Museum of Natural History in Paris. Berland wrote that "the substantial oeuvre that [Mello Leitão] left behind will keep his memory alive." "Ata da Sessão Ordinária da Academia Brasileira de Ciências, Homenagem Póstuma a Mello Leitão," July 12, 1949, Mello Leitão papers, ABC.
18. Roquette-Pinto to Francisco Campos, February 4, 1931, MN.BR.MN.DR.1932, SEMEAR.
19. Everton Coelho de Matos, "Brasil e Uruguai: Uma dívida que virou ponte" (Coursework, Pontifícia Universidade Católica do Rio Grande do Sul, Uruguaiana, Brazil, 2008), 9, 21–27, 33–38.
20. In 1921, Armanda Alberto founded an early elementary school called Escola Regional de Meriti. It had a library that was open to the public, served school lunch, and integrated families into the school. Alberto was one of the signers of the Manifesto of the Pioneers of New Education in 1932. She also took part in anticlerical movements alongside Edgard Süssekind de Mendonça, whom she married in 1928. The couple belonged to the Anticlerical League of Brazil (Liga Anticlerical do Brasil), whose organ was the anarchist newspaper *A Lanterna*. They were arrested in 1936 for involvement with communism. See José Damiro de Moraes, "Signatárias do Manifesto de 1932: Trajetórias e dilemas" (PhD diss., Universidade Estadual de Campinas, 2007), 149–223.
21. Mello Leitão, "O livro de minha vida," 26; "Intercambio intelectual con el Brasil," *La Mañana*, October 23, 1931.
22. "Conferencias," *El Imparcial*, October 30, 1931; "Nota," *El Diário*, October 23, 1931. On Fabre, see chap. 2 of this book. Maurice Maeterlinck (1862–1949), Belgian playwright and essayist, wrote best-selling novelized texts on bees, termites, and ants (*La vie des abeilles*, 1901; *La vie des termites*, 1927; and *La vie des fourmis*, 1930). He won the 1911 Nobel Prize for Literature.
23. On foreign policy under diplomat Rio Branco (1845–1912), who was minister of foreign affairs from 1902 to 1912, see Luís Claudio Vilafañe Santos, *O evangelho do Barão* (São Paulo: Editora Unesp, 2012).

24. "Un rato de amena conversación com el destacado intelectual brasileño," *El Imparcial*, November 4, 1931; "Con el profesor Mello Leitão, embajador de la intelectualidad brasileña," *El Debate*, October 28, 1931.
25. "La Conferencia de la Argentina, Brasil y Uruguay," *La Nación*, November 26, 1931. On Americanism during the early years of the Provisional Government, see Amado Luís Cervo and Clodoaldo Bueno, *A política externa brasileira, 1822–1985* (São Paulo: Ática, 1986), 71.
26. The paper's original title was "Zoogeografia de los opiliones sudamericanos." Mello Leitão, "O livro de minha vida," 27–28; "Hoy se realizara um acto em honor del aracnologo brasileño Dr. Mello Leitão," *La Prensa*, November 4, 1931; "Museu Nacional de Historia Natural ofrecerá hoy una recepción al doctor Mello Leitão," *La Nacion*, November 4, 1931.
27. Mello Leitão, "O livro de minha vida," 31.
28. Mello Leitão, "O livro de minha vida," 31–44.
29. "Nota informativa sobre la marcha de las tareas de impresión, II Reunion de Ciências Naturales," *Physis, Revista de la Asociación Argentina de Ciencias Naturales* 12 (1938): 376–77; "Acta de la sección Zoologia (Invertebrados)," *Physis, Revista de la Asociación Argentina de Ciencias Naturales* 17 (1939): xi-xx. The other Brazilians registered at the event were Pablo Campos Porto, Thomas Borgmeier, H. de Souza Lopes, J. F. de Teixeira, H. Lent, Fernando Milanes, A. Brade, R. Arlé, and Arthur Neiva.
30. Juan José Parodiz, "Cândido Firmino de Mello Leitão, 1886–1948," *Physis, Revista de la Asociación Argentina de Ciencias Naturales* 20:57 (1949): 205–6.
31. Letícia Pinheiro, *Política externa brasileira* (Rio de Janeiro: Zahar, 2004), 22–23. See also Francisco Corsi, "Política externa e desenvolvimento no Estado Novo," *Locus, Revista de História* 13:2 (2007): 253–56.
32. José Honório Rodrigues and Ricardo Seitenfus, *Uma história diplomática do Brasil, 1531–1945* (Rio de Janeiro: Civilização Brasileira, 1995), 347–89. Oswaldo Aranha's stance during the Vargas era was a complex one. He took active part in the Revolution of 1930, voiced his displeasure over the 1937 coup, and yet agreed to head the Ministry of Foreign Affairs. During his term, he had to deal with military sectors sympathetic to the Axis. See Stanley Hilton, *Oswaldo Aranha: Uma biografia* (Rio de Janeiro: Objetiva, 1994), 222–63; Frank McCann Jr., *The Brazilian-American Alliance 1937–1945* (Princeton: Princeton University Press, 1973): 49–76.
33. The paper that Mello Leitão submitted to the Seventh International Congress

of Entomology in Berlin is in "Notes sur les Proscopides,"in *Verhandlungen der VII International Kongress für Entomologie, du 15 au 20 août, 1938*, 292–302 (Berlin: Druck Von G. Uschmann, 1939). The paper did not focus on arachnids but on the bug commonly known as a stick insect; he described a new species that attacks the eucalyptus tree. See Carlos A. H. Flechtmann and Angelo L. T. Ottati, "*Tetanorhynchus leonardosi* (Mello-Leitão) (*Orthoptera: Proscopidae*), nova praga em eucaliptos," *Anais da Sociedade Entomológica do Brasil* 26:3 (1997): 583–87.

34. "Notas de crónica de la Segunda Reunión de Ciencias Naturales: Excursión a la bodega El Trapiche," *Physis, Revista de la Asociación Argentina de Ciencias Naturales* 17 (1939): 63; "Cinco minutos de conversación con la más alta autoridad em arañas," *La Libertad*, April 11, 1937.

35. The correspondence between Mello Leitão and Feio can be found in JF.O.MN.20, SEMEAR. On the shipment of specimens from Britain's Natural History Museum for classification by Mello Leitão, see Heloísa Alberto Torres to the British Ambassador, June 24, 1938, BR.MN.MN.DR.1939, SEMEAR.

36. Mello Leitão, "O livro de minha vida," 38–39. See also Clemente Estable, "Palabras de presentación, conferência del profesor doctor C. Mello Leitão," *Anales de la Faculdad de Medicina de Montevideo* 30:12 (1945): 947–50. See also "El profesor brasileño Dr. Cândido Mello Leitão, su próxima visita," *La Mañana*, August 23, 1945; "Llego ayer la mision cultural brasileña," *La Mañana*, September 4, 1945; "Inicia hoy su actuación la Embajada Cultural Brasileña," *La Razon*, September 5, 1945; "Reunión en el Museo de H. Natural em Homenaje al Prof. de Mello Leitão," *El Diário*, September 12, 1945.

37. On Clemente Estable (1894–1994), see www.histoemb.fmed.edu.uy/cestable.html and iibce.edu.uy/uas/cestable/estable.htm, accessed July 24, 2015.

38. "Agasajo anoche el circulo de la prensa a intelectuales brasileños," *La Nacion*, September 26, 1943; "Hubo un acto en el Museo Argentino de Ciencias Naturales," *La Nación*, September 27, 1945; Aloysio Mello Leitão to Feio, October 1945, JF.O.MN.2.18/3, SEMEAR.

39. Cervo and Bueno, *A política externa brasileira*, 72. See also Renato Lemos, "Política externa," in *Getúlio Vargas e seu tempo*, ed. Raul Silva Mendes, Paulo Capachuz, and Sergio Lamarão, 75–82 (Rio de Janeiro: BNDES, 2004).

40. Ana Luiza Beraba, *América aracnídea: Teias culturais interamericanas* (São Paulo: Civilização Brasileira, 2008). On the press under the Estado Novo, see Maria Helena Capelato, "Propaganda política e controle dos meios de comunicação,"

in *Repensando o Estado Novo*, ed. Dulce Pandolfi, 167–78 (Rio de Janeiro: FGV, 1999); Livia Lopes Neves, "Pensamento da América: Intelectualidade e Estado Novo em um projeto comungado (1941–1945)" (MA thesis, Universidade Federal de Santa Catarina, Florianópolis), 2013.

41. Beraba, *América aracnídea*, 146, 153.
42. Beraba, *América aracnídea*, 152. Here we have an example of the ambiguous nature of Vargas's foreign policy. Despite the opportunity for Latin American countries to exchange films through the Buenos Aires conference, it is interesting to remember that this same year (1936), Minister of Education Capanema sent Roquette-Pinto to Europe to assess film production there, and the scientist made important stops in Italy and Germany, as mentioned earlier.
43. Beraba, *América aracnídea*, 165–66.
44. See the following papers by Mello Leitão: "Notas sobre aracnídeos argentinos," *Anais da Academia Brasileira de Ciências* 3:2 (1931): 83–97; "Some Spiders of British Guiana Taken by Mr. C. W. Richards from the Nests of Solitary Wasps," *Anais da Academia Brasileira de Ciências* 11:2 (1939): 105–12; "Spiders of the Guiana Forest Collected by C. W. Richards," *Arquivos de Zoologia do Estado de São Paulo* 2 (1941): 175–97; "Contribuição ao conhecimento da fauna aracneológica das Guianas," *Anais da Academia Brasileira de Ciências* 20:2 (1948): 151–96; "Mais alguns novos opiliões sul-americanos," *Anais da Academia Brasileira de Ciências* 12:2 (1940): 93–107; "Um solífugo da Argentina e alguns opiliões da Colômbia," *Anais da Academia Brasileira de Ciências* 12:4 (1940): 301–11; Mello Leitão, "Notes on Peruvian Harvest Spiders," *Anais da Academia Brasileira de Ciências* 13:4 (1941): 319–22; "Novos solífugos do Chile e do México," *Anais da Academia Brasileira de Ciências* 14:4 (1942): 305–14; "Alguns opiliões novos da Colômbia," *Anais da Academia Brasileira de Ciências* 113:3 (1941): 165–71; "Catálogo das aranhas da Colômbia," *Anais da Academia Brasileira de Ciências* 13:4 (1941): 233–300; "Aracnidos recogidos en el Ecuador y el Peru por la Señora H. E. Frizell Don," *Comunicaciones zoológicas del Museo de Historia Natural de Montevideo* 5:1 (1943): 1–8; "Nuevos aracnidos sudamericanos de las colecciones del Museo de Historia Natural de Montevideo," *Comunicaciones zoologicas del Museu de História Natural de Montevideo* 21:1 (1944): 1–4; "Considerações sobre o gênero *Eusarcus* Perty e descrição de quatro novos Laniatores," *Anais da Academia Brasileira de Ciências* 17:2 (1945): 149–62; and with José Lacerda de Araújo Feio, "Notas sobre pequena coleção de aracnídeos do Peru," *Boletim do Museu Paraense Emílio Goeldi* 10 (1949): 131–324. Wolfgang Karl Weyrauch worked at the Estación Experimental Agrícola de Tingo María, but Mello

Leitão misidentified his affiliation as the "Instituto Agronômico de Tingo-Maria," of which I could find no record.

45. Mello Leitão, "O livro de minha vida," 52–69; Kury and Baptista, "Arachnological Papers," 1–17.
46. Charles Darwin, *On the Origin of Species* (Cambridge, MA: Harvard University Press, 2001), 346–410. On biogeography, see also Mayr, *The Growth of Biological Thought*, 439–54; Gould, *The Structure of Evolutionary Theory*, 113–15. Wallace's *Travels on the Amazon and Rio Negro* was translated into Portuguese in 1939 under the title *Viagens pelo Amazonas e rio Negro*, trans. Orlando Torres, Coleção Brasiliana 156 (São Paulo: Companhia Editora Nacional, 1939).
47. On the worldwide tendency to reject liberalism and liberal values in the 1930s, see Eric Hobsbawm, *The Age of Extremes: A History of the World 1914–1991* (New York: Vintage Books, 1996), 109–41.
48. Capelato, "O Estado Novo: O que trouxe de novo," 136–39; Antonio de Almeida Junior, "Do declínio do Estado Novo ao suicídio de Getúlio Vargas," in *O Brasil Republicano*, ed. Boris Fausto,, 227–29, Coleção História Geral da Civilização Brasileira 10 (São Paulo: Difel, 1983).
49. Mello Leitão, "Novos rumos da biogeografia," *Revista Brasileira de Geografia* 7:3 (1945): 464–66.
50. "Del profesor Mello Leitão," *La Mañana*, November 30, 1931; "Universitárias," *La Plata*, September 17, 1945.
51. Mello Leitão, *Glossário biológico*.
52. Mello Leitão, *Zoogeografia do Brasil*, 7–8.
53. Alfred Wegener, "The Origins of the Continents (1915)," trans. Wolfgang Jacoby, *Journal of Geodynamics* 32:1–2 (2001): 20–63; Patrick Hurley, "The Confirmation of Continental Drift," *Scientific American* 218 (1968): 52–64; Wolfgang Jacoby, "Modern Concepts of Earth Dynamics Anticipated by Alfred Wegener in 1912," *Geology: Journal of the Geological Society of America* 9:1 (1981): 25–27.
54. Mello Leitão, *Zoogeografia do Brasil*, 582–609. Mello Leitão based his checklists on James Peters, *Check List of Birds of the World* (Cambridge, MA: Harvard University Press, 1931). It was indeed trailblazing to view Wegener's theory with sympathetic eyes, as we see in Mello Leitão, "A gênese dos continentes e oceanos segundo Wegener," 49–54.
55. Moses was well known for his work in zoology applied to health and had published studies on *Taenia solium*, as discussed in Hélio A. G. Teive, Sérgio M.

De Almeida, and Lineu César Werneck, "The Brazilian Contribution to the Study of Neurocysticercosis: Moses and Lange's Role in Cerebrospinal Fluid Diagnosis," *Arquivos de Neuro-Psiquiatria* 64:2B (2006): 534–37.

56. Mello Leitão, "Discurso de posse na Academia Brasileira de Ciências," Mello Leitão papers, ABC.

57. Alexander Petrunkevitch (1875–1964) was a Russian immigrant to the United States and an eminent arachnologist at the Yale Peabody Museum of Natural History, where he worked until his death. See www.peabody.yale.edu/archives/ypmbios/petrunkevitch.html, accessed November 14, 2009. Clarence Hoffman was an entomologist with the U.S. Department of Agriculture. The speech given by Moses at Mello Leitão's inauguration was published in a newspaper, but the clipping indicates neither place nor date of publication; it can be found in MN.BR.MR JF.O.DR.3, SEMEAR.

58. This point is defended by Sá, *A ciência como profissão*, 73–87.

59. Mello Leitão, "Discurso de posse."

60. This takes us back to the overriding argument of chapter 1, i.e., that biological knowledge assumed strategic importance in the historical construction of Brazilian society in the early decades of the twentieth century, configuring a true "age of biology."

61. Cândido de Mello Leitão to Heitor Grillo, August 18, 1939, CO.1302 (Costa Lima papers), SEMEAR.

62. The following are textbooks by Mello Leitão: *Elementos de Zoologia de accordo com os programmas do Collegio Pedro II e da Escola Normal do Rio de Janeiro* (Rio de Janeiro: Francisco Alves, 1917); *Reprodução dos animais* (Rio de Janeiro: Francisco Alves, 1923); *Compêndio de botânica* (Rio de Janeiro: Francisco Alves, 1924); *Compêndio de zoologia* (Rio de Janeiro: Francisco Alves, 1924); *Noções de biologia geral* (Rio de Janeiro: Francisco Alves, 1930); *Curso elementar de história natural*, 4 vols. Série Livros Didáticos 15, 22, 36, 42 (São Paulo: Companhia Editora Nacional, 1933–35); *Biologia geral* (São Paulo: Companhia Editora Nacional, 1940); *Compêndio Brasileiro de Biologia*, 2 vols. (São Paulo: Companhia Editora Nacional, 1942); *Glossário biológico*; and with Pecegueiro do Amaral, *Noções de ciências naturais para a quarta série do ginasial*, 2 vols. (São Paulo: Companhia Editora Nacional, 1942), and *Noções de física, química e biologia para os cursos de comércio* (São Paulo: Companhia Editora Nacional, 1945).

63. On São Paulo, see Luciana Viviani, *A biologia necessária, formação de professoras e escola normal* (Belo Horizonte: Argvmentvm, 2007), 20–24, 116–17, 236,

257–58. On Rio de Janeiro, see Vidal, *O exercício disciplinado do olhar*. The question of the history of biology in school curricula is a complicated one, demanding its own study. In 1930, Mello Leitão lauded the inclusion of general biology in the curriculum at the Rio de Janeiro Normal School. See *Noções de biologia geral*, 5.

64. Mariana Cassab and Sandra Scovedo Selles, "Investigando os rumos escolares da disciplina História Natural no Colégio Pedro II: As atas de concursos para professores como fonte histórica," *Revista Contemporânea de Educação* 3:6 (2008): 238–58; Viviani, *A biologia necessária*, 7–9.

65. Mello Leitão, *Ainda em defesa de um parecer: Documentos sobre a desaprovação de um compendio de zoologia pela Câmara Nacional do Livro Didático* (Rio de Janeiro: Editora Jornal do Commércio, 1946); Waldemiro Potsch, *Novos erros, novas cópias, novos plágios do prof. Cândido de Mello Leitão* (Rio de Janeiro: Gráfica Editora Aurora, 1947). See also Maria Cristina Ferreira dos Santos, "A biologia de Cândido de Mello Leitão e a História Natural de Waldemiro Potsch: Professores, autores e livros didáticos, conhecimento e poder em disputa na constituição da biologia escolar" (PhD diss., Universidade Federal Fluminense, Niterói, 2013).

66. Mello Leitão, "Comentários sobre los araneísmos: Conferencia en la Misión Cultural Brasileña, pronunciada en la Faculdad de Medicina de Montevideo," *Anales de la Facultad de Medicina de Montevideo* 30:12 (1945): 964.

67. On Dobzhansky, his role in the revival of Darwinism, and his participation in Brazilian scientific circles, see Thomas Glick, "A Fundação Rockefeller e a emergência da genética no Brasil (1943–1960)," in *A recepção do darwinismo no Brasil*, 145–74.

68. "General Information: Conference for the Establishment of the IUPN," Paris, July 20, 1948, NS/UIPN/1 UNESDOC Database, http://unesdoc.unesco.org/images/0015/001547/154739eb.pdf, accessed July 25, 2015.

69. "Liste des organismes nationaux et internationaux non gouvernementaux invités à la Conférence de Fontainebleau par l'Union Internationale Provisoire pour la Protection de la Nature," Paris, July 19, 1948, 2, NS/UIPN/5B, UNESDOC Database, http://unesdoc.unesco.org/images/0015/001547/154745mb.pdf, accessed July 25, 2015.

70. Charles J. Bernard, ed., *International Union for the Protection of Nature* (Brussels: IUCN Library, 1948), 6–7, 23–26; "Procès verbal provisoire de la quatrième séance tenue au Palais de Fontainebleau," Paris, January 6, 1949, 5, NS/UIPN/TECH.

SYMP/SR.4 UNESDOC Database, http://unesdoc.unesco.org/images/0015/001547/154738fb.pdf, accessed July 25, 2015.

CONCLUSION

1. Eschewing dichotomous perspectives, some more recent historians have developed new approaches to the study of the scientific trajectories of individuals. See, e.g., "Scientific Personae and Their Histories," ed. Lorraine Daston and H. Otto Sibum, special issue, *Science and Context* 16:1–2 (2003): 1–269.
2. Bloch, *The Historian's Craft*, 31.
3. Bloch, *The Historian's Craft*, 25.

PORTUGUESE AND ENGLISH NAMES OF INSTITUTIONS AND EVENTS

ARGENTINA ENTOMOLOGICAL SOCIETY: Sociedad Entomológica Argentina

ARGENTINA MUSEUM OF NATURAL SCIENCES: Museo Argentino de Ciencias Naturales (now the Museo Argentino de Ciencias Naturales Bernardino Rivadávia)

ASSISTANCE SERVICE FOR THE TEACHING OF NATURAL HISTORY: Serviço de Assistência ao Ensino de História Natural

BERNARDINO RIVADÁVIA MUSEUM OF NATURAL HISTORY: Museo Argentino de Ciencias Naturales Bernardino Rivadávia

BIOLOGICAL INSTITUTE OF AGRICULTURAL AND ANIMAL DEFENSE: Instituto Biológico de Defesa Agrícola e Animal

BRAZILIAN ACADEMY OF LETTERS: Academia Brasileira de Letras (ABL)

BRAZILIAN ACADEMY OF SCIENCES: Academia Brasileira de Ciências (ABC)

BRAZILIAN COMMUNIST PARTY: Partido Comunista Brasileiro (PCB)

BRAZILIAN COUNCIL ON THE OVERSIGHT OF ARTISTIC AND SCIENTIFIC EXPEDITIONS: Conselho de Fiscalização das Expedições Artísticas e Científicas do Brasil

BRAZILIAN EDUCATION ASSOCIATION: Associação Brasileira de Educação (ABE)

BRAZILIAN EUGENICS CONGRESS: Congresso Brasileiro de Eugenia

BRAZILIAN EXPEDITIONARY FORCE: Força Expedicionária Brasileira (FEB)

BRAZILIAN FILM ASSOCIATION: Associação Brasileira de Cinema

BRAZILIAN HISTORICAL AND GEOGRAPHICAL INSTITUTE: Instituto Histórico e Geográfico Brasileiro (IHGB)

CAJURU BIOLOGICAL STATION: Estação Biológica do Cajuru

CENSORSHIP COMMISSION: Comissão de Censura

CHILDREN'S POLYCLINIC: Policlínica de Crianças

CHILEAN ACADEMY OF NATURAL SCIENCES: Academia Chilena de Ciencias Naturales

COLOMBIAN ACADEMY OF EXACT, PHYSICAL, AND NATURAL SCIENCES: Academia Colombiana de Ciencias Exactas, Físicas y Naturales

COMMISSION FOR THE CENSORSHIP AND SELECTION OF EDUCATIONAL FILMS: Comissão de Censura e Seleção de Filmes Educativos

CONGRESS ON THE HISTORY OF BRAZIL: Congresso de História do Brasil

CONSTITUTIONALIST REVOLUTION: Revolução Constitucionalista

DEPARTMENT OF PROPAGANDA AND CULTURAL PROMOTION: Departamento de Propaganda e Difusão Cultural (DPDC)

DIRECTORATE OF INFORMATION, STATISTICS, AND DISSEMINATION: Diretoria de Informações, Estatística e Divulgação

DOM VITAL CENTER: Centro Dom Vital

FEDERAL FOREST COUNCIL: Conselho Federal Florestal

FEDERAL REGISTER: Diário Oficial

FEDERAL SERUM THERAPY INSTITUTE (RIO DE JANEIRO): Instituto Soroterápico Federal

FEDERAL UNIVERSITY OF MARANHÃO: Universidade Federal do Maranhão

FEDERAL UNIVERSITY OF RIO DE JANEIRO: Universidade Federal do Rio de Janeiro

FILM TAX: Taxa Cinematográfica

FIRST AMERICAN CONGRESS OF NATIONAL COMMISSIONS FOR INTELLECTUAL COOPERATION: Primera Conferencia Americana de Comisiones Nacionales de Cooperación Intelectual

FIRST BRAZILIAN CONGRESS FOR THE PROTECTION OF NATURE: Primeira Conferência Brasileira de Proteção à Natureza

FOREST CODE: Código Florestal

FRIENDS OF NATURE CLUBS AT SCHOOLS: Clubes Escolares de Amigos da Natureza

GAME AND FISH CODE: Código de Caça e Pesca

GENERAL DIRECTORATE OF PUBLIC INSTRUCTION: Diretoria Geral da Instrução Pública

GREATER UNIVERSITY OF THE REPUBLIC (URUGUAY): Universidad Mayor de la República (now the Universidad de la República)
HAHNEMANNIAN INSTITUTE SCHOOL OF MEDICINE AND SURGERY: Escola de Medicina e Cirurgia do Instituto Hahnemanniano
HIGHER SCHOOL OF AGRICULTURAL SCIENCE AND VETERINARY MEDICINE: Escola Superior de Agricultura e Medicina Veterinária (now the Universidade Federal Rural do Rio de Janeiro)
INDIAN PROTECTION SERVICE: Serviço de Proteção aos Índios (SPI)
INTERNATIONAL EXHIBITION IN CELEBRATION OF THE CENTENNIAL OF INDEPENDENCE: Exposição Internacional do Centenário da Independência
LA PLATA MUSEUM OF NATURAL SCIENCES: Museo de La Plata—Facultad de Ciencias Naturales y Museo
LAW ON SCIENTIFIC EXPEDITIONS: Lei de Expedições Científicas
MANIFESTO OF THE PIONEERS OF NEW EDUCATION: Manifesto dos Pioneiros da Educação Nova
MENTAL HYGIENE LEAGUE: Liga de Higiene Mental
METEOROLOGICAL INSTITUTE: Instituto Meteorológico
MINE CODE: Código de Minas
MINEIROS MANIFESTO: Manifesto dos Mineiros
MINISTRY OF AGRICULTURE: Ministério da Agricultura
MINISTRY OF AGRICULTURE, INDUSTRY, AND COMMERCE: Ministério da Agricultura, Indústria e Comércio
MINISTRY OF EDUCATION AND PUBLIC HEALTH: Ministério da Educação e Saúde Pública (MESP)
MINISTRY OF FOREIGN AFFAIRS: Ministério das Relações Exteriores
MINISTRY OF JUSTICE AND INTERNAL AFFAIRS: Ministério da Justiça e Negócios Interiores
MINISTRY OF TRANSPORTATION AND PUBLIC WORKS: Ministério da Viação e Obras Públicas
NATIONAL ACADEMY OF ART: Academia Nacional de Belas Artes
NATIONAL ACADEMY OF MEDICINE: Academia Nacional de Medicina (ANM)
NATIONAL ACADEMY OF SCIENCES (CÓRDOBA, ARGENTINA): Academia Nacional de Ciencias
NATIONAL COMMISSION FOR THE PROTECTION OF SOUTH AMERICAN FAUNA: Comissão Nacional para a Proteção da Fauna Sul-Americana
NATIONAL GAME AND FISH COUNCIL: Conselho Nacional de Caça e Pesca

NATIONAL INSTITUTE OF EDUCATIONAL CINEMA: Instituto Nacional de Cinema Educativo (INCE)

NATIONAL LIBERATION ALLIANCE: Aliança Nacional Libertadora (ANL)

NATIONAL MUSEUM: Museu Nacional

NATIONAL MUSEUM OF NATURAL HISTORY (PARIS): Muséum National d'Histoire Naturelle

NATIONAL MUSEUM OF NATURAL HISTORY (URUGUAY): Museo Nacional de Historia Natural

NATIONAL SCHOOL OF AGRONOMY: Escola Nacional de Agronomia (now part of the Universidade Federal Rural do Rio de Janeiro)

NATIONAL SCHOOL OF PHILOSOPHY (RIO DE JANEIRO): Faculdade Nacional de Filosofia

NATIONAL SECURITY LAW: Lei de Segurança Nacional

NATIONAL STUDENT UNION: União Nacional dos Estudantes (UNE)

NATIONAL TEXTBOOK COMMISSION: Comissão Nacional do Livro Didático

NATIONAL UNIVERSITY OF LA PLATA: Universidad Nacional de La Plata

NATURAL HISTORY MUSEUM OF BARCELONA: Museu de Ciències Naturals de Barcelona

NATURAL HISTORY MUSEUM OF BASEL: Naturhistorisches Museum Basel

NITERÓI NORMAL SCHOOL: Escola Normal de Niterói

OSWALDO CRUZ INSTITUTE: Instituto Oswaldo Cruz

PARÁ MUSEUM OF NATURAL HISTORY AND ETHNOGRAPHY: Museu Paraense de História Natural e Etnografia

PAULISTA MUSEUM: Museu Paulista

POLYTECHNIC SCHOOL: Escola Politécnica

PRESS AND PROPAGANDA DEPARTMENT: Departamento de Imprensa e Propaganda (DIP)

PROF. MELLO LEITÃO BIOLOGY MUSEUM: Museu de Biologia Prof. Mello Leitão

PRO-SANITATION LEAGUE OF BRAZIL: Liga Pró-Saneamento do Brasil

PROVISIONAL GOVERNMENT: Governo Provisório

RIO DE JANEIRO INSTITUTE OF EDUCATION: Instituto de Educação do Rio de Janeiro

RIO DE JANEIRO NORMAL SCHOOL: Escola Normal do Rio de Janeiro

RIO DE JANEIRO SCHOOL OF MEDICINE: Faculdade de Medicina do Rio de Janeiro

ROYAL MUSEUM: Museu Real

SÃO PAULO STATE BACTERIOLOGICAL INSTITUTE: Instituto Bacteriológico do Estado de São Paulo (now the Instituto Adolfo Lutz)

SCHOOL OF EXACT, PHYSICAL, AND NATURAL SCIENCES (UNIVERSITY OF BUENOS AIRES): Facultad de Ciencias Exactas, Físicas y Naturales (now the Facultad de Ciencias Exactas y Naturales)

SCHOOL OF MEDICAL SCIENCES (PARAGUAY): Faculdad de Ciencias Médicas, Universidad Nacional de Asuncion

SCHOOL OF MEDICINE (URUGUAY): Faculdad de Medicina, Universidad de la República

SCHOOL OF PHILOSOPHY, SCIENCES, AND LITERATURE (UNIVERSITY OF SÃO PAULO): Faculdade de Filosofia, Ciências e Letras

SECOND MEETING OF THE NATURAL SCIENCES: II Reunión de Ciencias Naturales

SERUM THERAPEUTICS INSTITUTE (SÃO PAULO): Instituto Serumtherápico (now the Instituto Butatan)

SOCIETY OF THE FRIENDS OF ALBERTO TORRES: Sociedade dos Amigos de Alberto Torres

SOCIETY OF THE FRIENDS OF THE NATIONAL MUSEUM: Sociedade dos Amigos do Museu Nacional

SOCIETY OF THE FRIENDS OF TREES: Sociedade dos Amigos das Árvores

STRATEGIC TELEGRAPH COMMISSION OF MATO GROSSO TO AMAZONAS: Comissão de Linhas Telegráficas Estratégicas de Mato Grosso ao Amazonas (CLTEMTA)

UNIVERSITY OF BRAZIL: Universidade do Brasil

UNIVERSITY OF BUENOS AIRES: Universidad de Buenos Aires

UNIVERSITY OF HAMBURG: Universität Hamburg

UNIVERSITY OF MINAS GERAIS: Universidade de Minas Gerais (now the Universidade Federal de Minas Gerais)

UNIVERSITY OF MONTEVIDEO: Universidad de Montevideo

UNIVERSITY OF PARANÁ: Universidade de Paraná (now the Universidade Federal de Paraná)

UNIVERSITY OF RIO DE JANEIRO: Universidade do Rio de Janeiro (now the Universidade Federal do Rio de Janeiro)

UNIVERSITY OF SÃO PAULO: Universidade de São Paulo (USP)

UNIVERSITY OF THE FEDERAL DISTRICT: Universidade do Distrito Federal

URUGUAYAN HISTORICAL AND GEOGRAPHICAL INSTITUTE: Instituto Histórico y Geográfico del Uruguay

VACCINOGEN INSTITUTE: Instituto Vacinogênico

VALPARAISO SCIENTIFIC SOCIETY: Sociedad Científica de Valparaíso

WATER CODE: Código das Águas

BIBLIOGRAPHY

ARCHIVES

Academia Brasileira de Ciências, Rio de Janeiro, Brazil—ABC

Arquivo Edgard Leuenroth, Campinas, Brazil—AEL

Arquivo Público Mineiro, Belo Horizonte, Brazil—APM

Biblioteca da Universidade Federal de Minas Gerais, Belo Horizonte, Brazil—BU/UFMG

Biblioteca del Congreso de la Nación, Buenos Aires, Argentina—BCN

Biblioteca do Museu Nacional/UFRJ, Rio de Janeiro, Brasil—BMN

Biblioteca "Florentino Ameghino," Faculdad de Ciencias Naturales y Museo, Universidad Nacional de La Plata, La Plata, Argentina—BFA

Biblioteca Nacional de Uruguay

Biblioteca Nacional "Mariano Moreno," Buenos Aires, Argentina

Biblioteca Pública Estadual Luiz de Bessa, Belo Horizonte, Brazil

Fundação Biblioteca Nacional, Rio de Janeiro, Brazil—FBN

Hemeroteca da Biblioteca Pública Luiz de Bessa, Belo Horizonte, Brasil

Museo Argentino de Ciencias Naturales Bernardino Rivadávia, Buenos Aires, Argentina—MACN

Museo Nacional de Historia Natural, Montevideo, Uruguay—MNHN

Seção de Memória e Arquivo do Museu Nacional, Universidade Federal do Rio de Janeiro, Brazil—SEMEAR

UNESDOC Database online—http://www.unesco.org/new/en/unesdoc-database/

NEWSPAPERS AND MAGAZINES

A Noite, Rio de Janeiro, Brazil
A Noite Ilustrada, Rio de Janeiro, Brazil
A Vanguarda, Rio de Janeiro, Brazil
Avante, Rio de Janeiro, Brazil
Chácaras e Quintais, Rio de Janeiro, Brazil
Diário de Notícias, Rio de Janeiro, Brazil
Diário Official da União, Rio de Janeiro, Brazil
El Debate, Montevideo, Uruguay
El Diário, Montevideo, Uruguay
Electron, Rio de Janeiro, Brazil
El Imparcial, Montevideo, Uruguay
Fon-Fon!, Rio de Janeiro, Brazil
Gazeta de Notícias, Rio de Janeiro, Brazil
Jornal do Brasil, Rio de Janeiro, Brazil
Jornal do Commércio, Rio de Janeiro, Brazil
La Libertad, Mendoza, Argentina
La Mañana, Montevideo, Uruguay
La Nación, Buenos Aires, Argentina
La Plata, La Plata, Argentina
La Prensa, Buenos Aires, Argentina
La Razón, Montevideo, Uruguay
O Estado de S. Paulo, Brazil
O Imparcial, Rio de Janeiro, Brazil
O Paiz, Rio de Janeiro, Brazil
Radio, Rio de Janeiro, Brazil
Revista do Brasil, São Paulo, Brazil
Revista Nacional de Educação, Rio de Janeiro, Brazil
Uiára, Rio de Janeiro, Brazil

PRIMARY SOURCES

"Acta de la sección Zoologia (Invertebrados)." *Physis, Revista de la Asociación Argentina de Ciencias Naturale* 17 (1939): xi–xiii.

Almeida, Maria G. R. "Utilinda brincando ou o ensino da leitura e da escrita ao alcance do leigo e da própria criança pela sonografia." *Revista Nacional de Educação* 1:10 (1933): 22–32.

Amaral, J. Pecegueiro do, and Cândido de Mello Leitão. *Noções de ciências naturais para a quarta série do ginasial.* 2 vols. São Paulo: Companhia Editora Nacional, 1942.

———. *Noções de física, química e biologia para os cursos de comércio.* São Paulo: Companhia Editora Nacional, 1945.

"A polarização horizontal das ondas curtas." *Electron* 1:9 (1926): 15.

Arruda, Bento. *Por campos e mattas.* São Paulo: Editora Monteiro Lobato, 1925.

"Ata da Sessão Ordinária da Academia Brasileira de Ciências, Homenagem Póstuma a Mello Leitão." July 12, 1949. Cândido de Mello Leitão papers, ABC.

Azevedo, Fernando et al. *Manifesto dos Pioneiros da Educação Nova: A reconstrução educacional no Brasil ao povo e ao governo.* São Paulo: Companhia Editora Nacional, 1932.

Bernard, Charles J., ed. *International Union for the Protection of Nature.* Brussels: IUCN Library, 1948.

Bouvier, Eugène Louis. "Formigas agricultoras." *Revista Nacional de Educação* 1:4 (1933): 36–45.

Campos, Francisco. "A *Revista Nacional de Educação*: Carta manuscrita reproduzida." *Revista Nacional de Educação* 1:1 (1933): 3.

Carvajal, Gaspar de, Alonso Rojas, and Cristobal Acuña. *Descobrimentos do Rio das Amazonas.* Trans. and ed. Cândido de Mello Leitão. Coleção Brasiliana 203. São Paulo: Companhia Editora Nacional, 1941.

Carvalho, Alberto de. *Manual do Caçador.* São Paulo: privately printed, 1924.

Castro, Bernardo de. *Tiro ao voo.* Rio de Janeiro: privately printed, 1925.

Childe, Alberto. "A leitura dos hieróglifos." *Revista Nacional de Educação* 1:2 (1932): 24.

Cunha, Heitor Pereira da. *Viagens e caçadas em Mato Grosso.* Rio de Janeiro: Francisco Alves, 1922.

Daus, Federico. "Prólogo." In Cândido de Mello Leitão, *La vida en la selva*, 7–13. Biblioteca de Autores Brasileños Traducidos al Castellano 12. Buenos Aires: Imprenta López, 1949.

Duque Estrada, Osório. "Compêndio de botânica." *Jornal do Brasil*, March 28, 1925.

Estable, Clemente. "Palabras de presentación, conferência del profesor doctor C. Mello Leitão." *Anales de la Facultad de Medicina de Montevideo* 30:12 (1945): 947–50.

"Estação de rádio escola no Distrito Federal." *Revista Nacional de Educação* 2:16–17 (1934): 44–51.

Estatutos: Sociedade dos Amigos do Museu Nacional. Rio de Janeiro: M.E.S. Serviço Gráfico, 1937. MN 500.1074 S 678, SEMEAR.

Fonseca, Alvaro. "Discurso inaugural na Sociedade Carioca de Educação." *Revista Nacional de Educação* 1:4 (1933): 1–3.

Goeldi, Emílio. *As aves do Brasil.* Rio de Janeiro: Francisco Alves, 1922.

———. "Destruição das garças e guarás." *Boletim do Museu Paraense* 2 (1898): 27–42.

Guerra Duval, Fernando. "Palestras sobre fotografia." *Revista Nacional de Educação* 1:4 (1933): 64–69.

Ihering, Hermann von. "Devastação e conservação das matas." *Revista do Museu Paulista* 8 (1911): 485–500.

———. "Necessidade de uma lei federal de caça e proteção das aves." *Revista do Museu Paulista* 3 (1902): 228–60.

———. "Proteção às aves." *Revista do Museu Paulista* 9 (1914): 316–32.

Lacerda, Maurício de. "O comedor de escorpiões, comensalismo, parasitismo, imperialismo." *O Imparcial*, August 9, 1935.

Leão, Múcio. "Registro literário: A vida maravilhosa dos animais." *Jornal do Brasil.* Newspaper clipping. BR.MN.JF.O.MN.DR.3, SEMEAR.

Leme, Alberto Betim Paes. "Relatório dos Trabalhos do Museu Nacional referentes ao anno de 1936, dirigido ao Exmo. Snr. Dr. Gustavo Capanema, January 17, 1937, 1." BR.MN.MN.DR, SEMEAR.

Lobato, José Bento Monteiro. *A onda verde.* São Paulo: Globo, 2008.

———. *Caçadas de Pedrinho.* Série Literatura Infantil 9. São Paulo: Companhia Editora Nacional, 1933.

———. *Mr. Slang e o Brasil, Problema Vital.* 12th ed. São Paulo: Brasiliense, 1968.

———. *Problema vital, Jeca Tatu e outros textos.* São Paulo: Editora Globo, 2010.

———. *Urupês.* 30th ed. São Paulo: Brasiliense, 1984.

Lopes, Raimundo. "Antropogeografia." *Revista Nacional de Educação* 1:11–12 (1933): 17–23.

Lourenço Filho, Manoel. *Introducção ao estudo da Escola Nova.* São Paulo: Melhoramentos, 1930.

Marcgrave, Jorge. *História Natural do Brasil.* Trans. Alberto José de Sampaio. São Paulo: Imprensa Oficial do Estado, 1942.

Mello Leitão, Cândido de. *A biologia no Brasil.* Coleção Brasiliana 99. São Paulo: Companhia Editora Nacional, 1937.

———. "A gênese dos continentes e oceanos segundo Wegener." *Revista Nacional de Educação* 2:15 (1933): 49–54.

———. *Ainda em defesa de um parecer: Documentos sobre a desaprovação de um compendio de zoologia pela Câmara Nacional do Livro Didático.* Rio de Janeiro: Editora Jornal do Commercio, 1946.

———. "Alguns gêneros e espécies novas de araneidos do Brasil." *Brotéria, Série Zoologia* 13:2–3 (1915): 129–42.

———. "Alguns opiliões novos da Colômbia." *Anais da Academia Brasileira de Ciências* 13:3 (1941): 165–71.

———. "A methodologia da história natural." *O Estado de S. Paulo*, July 1, 1927.

———. "Aracnidos recogidos en el Ecuador y el Peru por la Señora H. E. Frizell Don." *Comunicaciones zoológicas del Museo de Historia Natural de Montevideo* 5:1 (1943): 1–8.

———. *A vida maravilhosa dos animais*. Série Iniciação Científica 7. São Paulo: Companhia Editora Nacional, 1935; 2nd ed. 1945.

———. *A vida na selva*. Série Iniciação Científica 20. São Paulo: Companhia Editora Nacional, 1940.

———. *Biologia geral: De acordo com os programas dos cursos pré-universitários*. Série Livros Didáticos 42. São Paulo: Companhia Editora Nacional, 1940.

———. "Catálogo das aranhas da Colômbia." *Anais da Academia Brasileira de Ciências* 13:4 (1941): 233–300.

———. "Comentários sobre los araneísmos: Conferencia en la Mision Cultural Brasileña, pronunciada en la Faculdad de Medicina de Montevideo." *Anales de la Facultad de Medicina de Montevideo* 30:12 (1945): 964.

———. *Compêndio brasileiro de biologia*. 2 vols. São Paulo: Companhia Editora Nacional, 1942.

———. *Compêndio de botânica*. Rio de Janeiro: Francisco Alves, 1924.

———. *Compêndio de zoologia*. Rio de Janeiro: Francisco Alves, 1924.

———. "Considerações sobre o gênero *Eusarcus* Perty e descrição de quatro novos Laniatores." *Anais da Academia Brasileira de Ciências* 17:2 (1945): 149–62.

———. "Contribuição ao conhecimento da fauna aracneológica das Guianas." *Anais da Academia Brasileira de Ciências* 20:2 (1948): 151–96.

———. *Curso elementar de história natural*. 4 vols. Série Livros Didáticos 15, 22, 36, 42. São Paulo: Companhia Editora Nacional, 1933–35.

———. "Discurso de posse na Academia Brasileira de Ciências." 1943. Mello Leitão papers, ABC.

———. "Dostoievsky e a revolução russa." *O Imparcial*, June 22, 1922.

———. "Doutorandos de 1908." *Jornal do Commércio*, December 5, 1948.

———. *Elementos de Zoologia de accordo com os programmas do Collegio Pedro II e da Escola Normal do Rio de Janeiro*. Rio de Janeiro: Francisco Alves, 1917.

———. "Ensino universitário na Rússia." O *Imparcial*, March 17, 1922.

———. "Genética." *Revista Nacional de Educação* 2:16–17 (1934): 30–35.

———. *Glossário biológico: Pequeno dicionário de termos técnicos empregados em ciências biológicas, botânica, ecologia, genética, zoologia*. São Paulo: Companhia Editora Nacional, 1946.

———. *História das expedições científicas no Brasil*. Coleção Brasiliana 209. São Paulo: Cia Ed. Nacional, 1941.

———. *La vida en la selva*. Biblioteca de Autores Brasileños Traducidos al Castellano 12. Buenos Aires: Imprenta López, 1949.

———. "Mais alguns novos opiliões sul-americanos." *Anais da Academia Brasileira de Ciências* 13:2 (1940): 93–107.

———. *Noções de biologia geral*. Rio de Janeiro: Francisco Alves, 1930.

———. "Notas arachnologicas, notas à margem do catálogo-índice de Petrunkevitch." *Brotéria, série Zoologia* 13:2–3 (1915): 143–44.

———. "Notas sobre aracnídeos argentinos." *Anais da Academia Brasileira de Ciências* 3:2 (1931): 83–97.

———. "Notes on Peruvian Harvest Spiders." *Anais da Academia Brasileira de Ciências* 13:4 (1941): 319–22.

———. "Notes sur les Proscopides." In *Verhandlungen der VII International Kongress für Entomologie, du 15 au 20 août, 1938*, 292–302. Berlin: Druck Von G. Uschmann, 1939.

———. "Novos rumos da biogeografia." *Revista Brasileira de Geografia* 7:3 (1945): 445–72.

———. "Novos solífugos do Chile e do México." *Anais da Academia Brasileira de Ciências* 14:4 (1942): 305–14.

———. "Nuevos aracnidos sudamericanos de las colecciones del Museo de Historia Natural de Montevideo." *Comunicaciones Zoologicas del Museu de História Natural de Montevideo* 21:1 (1944): 1–4.

———. *O Brasil visto pelos ingleses*. Coleção Brasiliana 82. São Paulo: Companhia Editora Nacional, 1937.

———. "O estudo da história natural, 1–2." *O Imparcial*, April 11–12, 1923.

———. "O livro de minha vida para ser lido pelos meus netos." Manuscript, 1946–47, 61 pp. Mello Leitão papers, ABC.

———. "O mais velho dos poetas . . . " *O Imparcial.* Newspaper clipping. BR.MN.JF.O.MN.DR-2, SEMEAR.

———. "Papel educativo do Museu Nacional de História Natural." *Revista Nacional de Educação* 1:2 (1932): 96–98.

———. "Pela eugenia." *O Imparcial,* 1922. Newspaper clipping. BR.MN.JF.O.MN.DR.2, SEMEAR.

———. "Prefácio." In Henry Bates, *O naturalista no rio Amazonas.* Trans. Cândido de Mello Leitão. Coleção Brasiliana 237–37A. São Paulo: Companhia Editora Nacional, 1944.

———. "Protegendo a nossa fauna silvestre (Entrevista)." *A Noite,* February 1933. Newspaper clipping. MN.JF.O.MN.DR.3, SEMEAR.

———. *Reprodução dos animais.* Rio de Janeiro: Francisco Alves, 1923.

———. "Some Spiders of British Guiana Taken by Mr. C. W. Richards from the Nests of Solitary Wasps." *Anais da Academia Brasileira de Ciências* 11:2 (1939): 105–12.

———. "Spiders of the Guiana Forest Collected by C. W. Richards." *Arquivos de Zoologia do Estado de São Paulo* 2 (1941): 175–97.

———. "Teia da vida." *O Imparcial,* October 27, 1922.

———. "Um solífugo da Argentina e alguns opiliões da Colômbia." *Anais da Academia Brasileira de Ciências* 12:4 (1940): 301–11.

———. "Vida e obra científica de Alberto José de Sampaio e Afrânio Peixoto." *Jornal do Commércio,* December 16, 1947.

———. *Visitantes do Primeiro Império.* Coleção Brasiliana 32. São Paulo: Companhia Editora, 1934.

———. *Zoogeografia do Brasil.* Coleção Brasiliana 77. São Paulo: Companhia Editora Nacional, 1937; 2nd ed. rev., 1947.

Mello Leitão, Cândido de, and José Lacerda de Araújo Feio. "Notas sobre pequena coleção de aracnídeos do Peru." *Boletim do Museu Paraense Emílio Goeldi* 10 (1949): 131–324.

"Modelos para obter licença de installação de um apparelho receptor radiotelephonico." *Radio* 1:16 (1926): 3.

Monteiro Lobato, José Bento. See under Lobato.

Morize, Henrique. "Aparelho automático para a recepção do sinal S.O.S." *Radio* 23 (1924): 9.

Naken, Theodore. "Fotografia e reprodução do som." *Electron* 1:16 (1926): 7–8.

Neiva, Arthur. "Prefácio." In Eurico Santos, *Pássaros do Brasil,* 8–9. 2nd ed. Rio de Janeiro: F. Briguiet, 1948.

"Nota." *Boletim do Museu de Biologia Prof. Mello Leitão.* Mello Leitão papers, ABC.

"Nota informativa sobre la marcha de las tareas de impresión, II Reunion de Ciências Naturales." *Physis, Revista de la Asociación Argentina de Ciencias Naturales* 12 (1938): 376–77.

"Notas de crónica de la Segunda Reunión de Ciencias Naturales: Excursión a la bodega El Trapiche." *Physis, Revista de la Asociación Argentina de Ciencias Naturales* 17 (1939): 63.

"O céu do Brasil." *Revista Nacional de Educação* 4 (1933): 80–84.

"O mais simples e mais econômico aparelho receptor de galena." *Electron* 1:10 (1926): 10.

O Museu Nacional: Commemoração do primeiro centenário da Independência do Brasil, notas e informações. Rio de Janeiro: Imprensa Nacional, 1922.

"Os cursos da Rádio Sociedade." *Electron* 1:1 (1926): 7.

"Os cursos na Associação Brasileira de Educação: O professor Roquette-Pinto encerrou as suas conferências sobre anthropologia." *Electron* 1:14 (1926): 4.

"O terceiro aniversário da Rádio Sociedade do Rio de Janeiro." *Electron* 1:6 (1926): 4.

Parodiz, Juan José. "Cândido Firmino de Mello Leitão, 1886–1948." *Physis, Revista de la Asociación Argentina de Ciencias Naturales* 20:57 (1949): 205–6.

Peixoto, Afrânio. "Discurso pronunciado no banquete oferecido ao prof. Miguel Pereira, em 19 de maio de 1918." In *Afrânio versus Afrânio*, ed. P. Leão et al., 29–37. Niterói: Tipografia "Jeronimo Silva," 1922.

Pereira, Lucia Miguel. "Livros: A vida maravilhosa dos animais." *Gazeta de Notícias*, August 25, 1945.

Pinto, Mário Olivério. "Resultados ornithológicos de uma excursão pelo oeste de São Paulo e sul de Mato Grosso." *Revista do Museu Paulista* 17 (1932): 689–826.

Pires, Washington. "Discurso de posse na pasta da Educação e Saúde Pública." *Revista Nacional de Educação* 1:1 (1932): 1–2.

Potsch, Waldemiro. *Novos erros, novas cópias, novos plágios do prof. Cândido de Mello Leitão, minhas novas descobertas.* Rio de Janeiro: Gráfica Editora Aurora, 1947.

"Primeiras letras." *Revista Nacional de Educação* 1:2 (1932): 80.

"Projeto de Decreto para regulamentação da caça no território nacional, mandado publicar para conhecimento dos cidadãos que queiram apresentar ao Ministro da Educação sugestões a respeito, no prazo de três meses, contados da primeira publicação." *Diário Official da União*, November 22, 1932.

Roquette-Pinto, Edgard. "Alberto Torres." *Revista Nacional de Educação* 2:18–19 (1934): 1–6.

———. "Apresentação." *Uiára* 1:1 (1937): 3.

———. "Comentários." *Revista Nacional de Educação* 1:3 (1932): 1.

———. *Conceito actual da vida*. Rio de Janeiro: Süssekind de Mendonça, 1922.

———. *Ensaios brasilianos*. Coleção Brasiliana 190. São Paulo: Companhia Editora Nacional, 1941.

———. *Ensaios de anthropologia brasiliana*. Coleção Brasiliana 22. São Paulo: Companhia Editora Nacional, 1933.

———. "Euclydes da Cunha, naturalista." *Revista do Brasil* 8:29 (1918): 20–38.

———. "O Brasil e a anthropogeographia." *Revista do Brasil* 3:12 (1916): 322–35.

———. "O Brasil e a radiocultura." *Radio* 1:19 (1924): 9.

———. "O cinema e a educação popular no Brasil." *Revista Nacional de Educação* 1:5 (1933): 1–9.

———. "Relatório dos trabalhos do Museu Nacional em 1929." BR.MN.MN. DR.1930, SEMEAR.

———. "Relatório dos trabalhos do Museu Nacional em 1930." BR.MN.MN. DR.1931, SEMEAR.

———. "Relatório dos trabalhos do Museu Nacional em 1931." BR.MN.MN. DR.1932, SEMEAR.

———. *Rondônia*. 4th ed. Coleção Brasiliana 39. São Paulo: Companhia Editora Nacional, 1935.

———. "Seção de Assistência ao Ensino da História Natural do Museu Nacional, relatório de 1930." SAE 146–5, SEMEAR.

———. *Seixos rolados*. Rio de Janeiro: Mendonça, Machado, 1927.

———. "Serviço de Assistência ao Ensino da História Natural, relatório de 1929." SAE 146.5, Doc. 7.09A, folder 105, SEMEAR.

———. "Prefácio." In Mello Leitão, *A biologia no Brasil*, 9–12.

———. "Prefácio." In Sampaio, *Biogeographia dynamica*, 5–6.

———. "Rádio Educação do Brasil." *Electron* 1:6 (1926): 15.

Roquette-Pinto, Edgard, Alberto José de Sampaio, and Cândido de Mello Leitão. "Considerações gerais: Projeto de decreto regulamentando a caça silvestre." Anexo ao Ofício 195, October 21, 1932. BR.MN.MN.DR, SEMEAR.

Roquette-Pinto, Paulo. "O Museu Nacional." *Uiára* 1:1 (1937): 6.

Sampaio, Alberto José de. *A alimentação sertaneja e do interior da Amazônia: Onomástica da alimentação rural*. Coleção Brasiliana 238. São Paulo: Companhia Editora Nacional, 1944.

———. "A carnaúba." *Revista Nacional de Educação* 1:7 (1933): 54–59.

———. *Biogeographia dynamica: A natureza e o homem no Brasil*. Coleção Brasiliana 53. São Paulo: Companhia Editora Nacional, 1935.

———. "O babassu." *Revista Nacional de Educação* 1:6 (1933): 37–43.

———. *Phytogeographia do Brasil.* Coleção Brasiliana 35. São Paulo: Companhia Editora Nacional, 1934.

———, ed. "Primeira Conferência Brasileira de Proteção à Natureza." *Boletim do Museu Nacional* 11:1 (1935): 1–230.

———, ed. "Primeira Conferência Brasileira de Proteção à Natureza." *Boletim do Museu Nacional* 11:2 (1935): 1–116.

———, ed. *Programa da Primeira Conferência Brasileira de Proteção à Natureza.* Rio de Janeiro: Sociedade dos Amigos das Árvores, 1933.

Serrano, Jonathas. "Carta a Roquette-Pinto." *Revista Nacional de Educação* 1:8 (1933): 2–4.

———. *Como se ensina História.* São Paulo: Melhoramentos, 1935.

———. *Methodologia da História na aula primária.* Rio de Janeiro: Francisco Alves, 1917.

Sinzig, Pedro. *Cancioneiro de modinhas populares.* Rio de Janeiro: B. Herder, 1914.

———. "Como ouvir música." *Revista Nacional de Educação* 1:4 (1933): 11–14.

———. *O Brasil cantando.* Petropolis: Editora Vozes, 1938.

Taunay, Afonso. "Prefácio." In Mello Leitão, *Visitantes do Primeiro Império*, 9–11.

Teixeira, Anísio. "Instruções regulando a censura e seleções de filmes educativos." *Revista Nacional de Educação* 1:9 (1933): 19–29.

Teixeira de Freitas, Mário Augusto. "Estatística e educação." *Revista Nacional de Educação* 1:1 (1933): 56–59.

———. *O ensino primário no Brasil.* São Paulo: Melhoramentos, 1934.

———. *O que dizem os números sobre ensino primário.* São Paulo: Melhoramentos, 1937.

———. "Primeira ronda." *Revista Nacional de Educação* 2:11–12 (1933): 63–64.

Torres, Alberto. *A organização nacional.* Rio de Janeiro: Imprensa Nacional, 1914.

———. *O problema nacional brasileiro.* Rio de Janeiro: Imprensa Nacional, 1914.

Vargas, Getúlio. "Educar." *Revista Nacional de Educação* 1:11–12 (1933): 1–9.

Wallace, Alfred R. *Viagens pelo Amazonas e rio Negro.* Translated by Orlando Torres Coleção. Brasiliana 156. São Paulo: Companhia Editora Nacional, 1939.

SECONDARY SOURCES

Accácio, Liéte. "A Escola Normal do Rio de Janeiro: A construção da profissão docente." *Revista de Pedagogia* 2:5 (2009): 4.

Alberti, Samuel. "Objects and the Museum." *Isis* 96:4 (2005): 559–72.

Alberti, Verena. "O século do moderno: Modos de vida e consumo na República." In *A República no Brasil*, ed. Angela de Castro Gomes, Dulce Pandolfi, and Verena Alberti, 260–319. Rio de Janeiro: Nova Fronteira, Editora FGV, 2002.

Almeida, Marta de. "São Paulo na virada do século XX: Um laboratório de saúde pública para o Brasil." *Tempo* 10:19 (2005): 77–89.

Almeida Junior, Antonio Mendes de. "Do declínio do Estado Novo ao suicídio de Getúlio Vargas." In *O Brasil Republicano*, ed. Bóris Fausto, 226–55. 2nd ed. Coleção História Geral da Civilização Brasileira 10. São Paulo: Difel, 1983.

Alonso, Angela. *Ideias em movimento: A geração de 1870 na crise do Brasil Império*. São Paulo: Paz e Terra, 2002.

———. "O positivismo de Luís Pereira Barreto e o pensamento brasileiro no final do século XIX." São Paulo: Instituto de Estudos Avançados da Universidade de São Paulo, 1995. http://www.iea.usp.br/publicacoes/textos/alonsopositivismo .pdf, accessed October 17, 2009.

Alves, José Jerônimo de Alencar. "As ciências na Academia e as expectativas de progresso e modernização, Brasil 1916–1929." In *Espaços da ciência no Brasil 1800–1930*, ed. Maria Amélia Dantes, 185–202. Rio de Janeiro: Editora Fiocruz, 2001.

Andermann, Jen. "Espetáculos da diferença: A Exposição Antropológica Brasileira de 1882." In *Galerias del progreso*, ed. Beatriz González and Jen Andermann, 151–93. Buenos Aires: Beatriz Viterbo Editora, 2006.

———. *The Optic of the State*. Pittsburgh: University of Pittsburgh Press, 2007.

Aragão, Mario. "Angelo Moreira da Costa Lima: De candidato a cirurgião a entomologista." *Cadernos de Saúde Pública* 4:4 (1988): 353–55.

Azevedo, Carmen Lucia, Marcia Camargos, and Vladimir Saccheta. *Furacão na Botocúndia*. São Paulo: Senac, 1997.

Bamberger, Yael, and Tali Tal. "The Learning Environment of Natural History Museums: Multiple Ways to Capture Students' Views." *Learning Environments Research* 12 (2009): 115–29.

Barbosa, Marialva. *História Cultural da Imprensa*. Rio de Janeiro: Mauad, 2007.

Barbour, Michael G. "Ecological Fragmentation in the Fifties." In *Uncommon Ground: Rethinking the Human Place in Nature*, ed. William Cronon, 233–55. New York: W. W. Norton, 1996.

Barros, Geraldo Mendes. *Guilherme Guinle, ensaio biográfico*. Rio de Janeiro: Agir, 1982.

Benchimol, Jaime. "Adolpho Lutz: Um esboço biográfico." *História, Ciências, Saúde—Manguinhos* 10:1 (2003): 13–83.

———. "Domingos José Freire e os primórdios da bacteriologia no Brasil." *História, Ciências, Saúde—Manguinhos* 2:1 (1995): 67–98.

———. *Manguinhos, do sonho à vida: A ciência na Belle Époque.* Rio de Janeiro: Editora Fiocruz, 1990.

Benchimol, Jaime, and Magali Romero Sá, eds. *Adolpho Lutz, obra completa.* Rio de Janeiro: Editora Fiocruz, 2006.

Benchimol, Jaime, and Luiz Antonio Teixeira. *Cobras, lagartos e outros bichos: Uma história comparada dos institutos Oswaldo Cruz e Butantan.* Rio de Janeiro: Editora Fiocruz, 1993.

Beraba, Ana Luiza. *América aracnídea: Teias culturais interamericanas.* São Paulo: Civilização Brasileira, 2008.

Bittencourt, José Neves. "Território largo e profundo: Os acervos dos museus do Rio de Janeiro como representação do Estado Imperial 1808–1889." PhD diss., Universidade Federal Fluminense, Niterói, Brazil, 1997.

Bloch, Marc. *The Historian's Craft.* New York: Vintage, 1964.

Bloch, Pedro. *Vida e obra de Angelo Moreira da Costa Lima.* Coleção Vultos da Ciência Brasileira 1. Rio de Janeiro: Conselho Nacional de Pesquisas, 1968.

Bomeny, Helena. "Novos talentos, vícios antigos: Os renovadores e a política educacional." *Estudos Históricos* 6:11 (1993): 24–39.

Boucher, François. *A History of Costume in the West.* New York: Thames and Hudson, 1987.

Bowler, Peter J. *The Earth Encompassed.* New York: Norton, 1993.

Brinkmann, Sören. "Milk and Modernity: Ideology and Nutrition Policies during the Vargas Era." *História, Ciências, Saúde—Manguinhos* 21:1 (2014): 263–80.

Britto, Nara. *Oswaldo Cruz: A construção de um mito na ciência brasileira.* Rio de Janeiro: Editora Fiocruz, 1995.

Bulcão, Lucia Grando, Almir Chaiban El-Kareh, and Jane Dutra Sayd. "Ciência e ensino médico no Brasil." *História, Ciências, Saúde—Manguinhos* 14:2 (2007): 469–87.

Capanema, Carolina. "A natureza no projeto de construção de um Brasil moderno e a obra de Alberto José de Sampaio." MA thesis, Universidade Federal de Minas Gerais, 2006.

Capel, Horacio. "El asociacionismo científico en Iberoamerica." *Interciencias* 17:3 (1992): 168–76.

Capelato, Maria Helena. "O Estado Novo: O que trouxe de novo?" In *O Brasil Republicano: O tempo do nacional-estatismo*, ed. Jorge Ferreira and Lucília Delgado, 107–43. Rio de Janeiro: Civilização Brasileira, 2003.

———. "Propaganda política e controle dos meios de comunicação." In *Repensando o Estado Novo*, ed. Dulce Pandolfi, 167–78. Rio de Janeiro: Editora FGV, 1999.

Caron, Joseph A. "Biology in the Life Sciences: A Historiographical Contribution." *History of Science* 26 (1988): 223–68.

Carvalho, José Murilo de. *Os bestializados*. São Paulo: Companhia das Letras, 1987.

Carvalho, Marta Maria Chagas de. *Molde nacional e fôrma cívica*. Bragança Paulista: Editora USF, 1998.

Carvalho, Marta Maria Chagas de, and Maria Rita de Almeida Toledo. "Reforma escolar, pedagogia da Escola Nova e usos do impresso." *Contemporaneidade e Educação* 5:7 (2000): 71–91.

Cassab, Mariana, and Sandra Scovedo Selles. "Investigando os rumos escolares da disciplina História Natural no Colégio Pedro II: As atas de concursos para professores como fonte histórica." *Revista Contemporânea de Educação* 3:6 (2008): 238–58.

Cervo, Amado Luís, and Clodoado Bueno. *A política externa brasileira, 1822–1985*. São Paulo: Ática, 1986.

Chalhoub, Sidney. *Cidade febril: Cortiços e epidemias na Corte Imperial*. São Paulo: Companhia das Letras, 1996.

———. *Trabalho, lar e botequim*. Campinas: Editora da Unicamp, 2001.

Chauí, Marilena. "História a contrapelo." In Edgard De Decca, *1930: O silêncio dos vencidos*, 11–30. São Paulo: Brasiliense, 1981.

Corrêa, Edison, and Sebastião Gusmão. *85 anos da Faculdade de Medicina da UFMG*. Belo Horizonte: Coopmed, 1996.

Corsi, Francisco Luiz. "Política externa e desenvolvimento no Estado Novo." *Locus, Revista de História* 13:2 (2007): 253–56.

Darwin, Charles. *On the Origin of Species*. Cambridge, MA: Harvard University Press, 2001.

Dean, Warren. *Brazil and the Struggle for Rubber*. Cambridge: Cambridge University Press, 2002.

———. *With Broadax and Firebrand: Destruction of the Brazilian Atlantic Forest*. Berkeley: University of California Press, 1997.

De Decca, Edgard. *1930: O silêncio dos vencidos*. São Paulo: Brasiliense, 1981.

Diacon, Todd. *Stringing Together a Nation: Cândido Mariano da Silva Rondon and the Construction of a Modern Brazil, 1906–1930*. Durham: Duke University Press, 2004.

Dias, Maria Odila Leite. "A interiorização da metrópole." In *1822: Dimensões*, ed. Carlos Guilherme Motta, 160–84. São Paulo: Perspectiva, 1972.

Dicionário histórico-biográfico das ciências da saúde no Brasil (1832–1930). Rio de Janeiro: Fundação Oswaldo Cruz. www.dichistoriasaude.coc.fiocruz.br.

Domingues, Heloisa Maria Bertol. "As demandas científicas e a participação do Brasil nas Exposições Internacionais." *Quipu* 12:2 (1999): 203–15.

———. "Controvérsias evolucionistas no Brasil do século XIX." In *A recepção do darwinismo no Brasil*, ed. Heloisa Maria Bertol Domingues, 97–123. Rio de Janeiro: Editora Fiocruz, 2003.

———. "Heloísa Alberto Torres e o inquérito nacional sobre ciências naturais e antropológicas, 1946." *Boletim do Museu Paraense Emílio Goeldi, Ciências Humanas* 5:3 (2010): 625–43.

Domingues, Heloisa Maria Bertol, and Magali Romero Sá. "O museu nacional e o ensino das ciências naturais no Brasil no século XIX." *Revista da Sociedade Brasileira para o Progresso da Ciência* 15 (1996): 79–87.

Domingues, Ivan, ed. *Conhecimento e transdisciplinaridade*. Belo Horizonte: Editora UFMG, 2005.

Drayton, Richard. *Nature's Government: Science, Imperial Britain, and the "Improvement" of the World*. New Haven: Yale University Press, 2000.

Drummond, José Augusto. "A legislação ambiental brasileira de 1934 a 1988." *Ambiente e Sociedade* 2:3–4 (1998–99): 127–49.

Duarte, Regina Horta. "Between the National and the Universal: Natural History Networks in Latin America in the Nineteenth and Twentieth Centuries." *Isis* 104:4 (2013): 777–87.

———. "Biologia e sociedade no Brasil dos anos 1930: Práticas de escrita e divulgação científica em Cândido de Mello Leitão." In *Ciência, história e teoria*, ed. Betânia Gonçalves Figueiredo and Mauro Lúcio Leitão Condé, 13–40. Belo Horizonte: Argvmentvm, 2005.

———. "Biologia, natureza e República no Brasil nos escritos de Mello Leitão (1922–1945)." *Revista Brasileira de História* 29:58 (2009): 321–44.

———. "Em todos os lares, o conforto moral da ciência e da arte: A Revista Nacional de Educação e a divulgação científica no Brasil (1932–1934)." *História, Ciências, Saúde—Manguinhos* 11:1 (2004): 33–56.

———. "Entre fábulas y arcanos: Proyecciones políticas y sociales del pensamiento biológico sobre la naturaleza en Brasil, 1922–1937." In *Naturaleza en declive: Miradas a la historia ambiental de América Latina y el Caribe*, ed. Reinaldo Funes, 317–46. Valencia: Centro Francisco Tomás y Valiente, UNED; Alzira-Valencia: Fundación de Historia Social, 2008.

———. "História e biologia: Diálogos possíveis, distâncias necessárias." *História, Ciências, Saúde—Manguinhos* 16:4 (2009): 927–40.

———. "Limites e fronteiras entre história e biologia em Michel Foucault: As palavras e as coisas e o surgimento da biologia no século XIX." In *Cartografias de Foucault*, ed. Durval Albuquerque Junior, Alfredo Veiga-Neto, and Alipio Souza Filho, 343–54. Belo Horizonte: Autêntica, 2008.

———. "Pássaros e cientistas no Brasil: Em busca de proteção, 1894–1938." *Latin American Research Review* 41:1 (2006): 3–26.

Dugatkin, Lee A. *Mr. Jefferson and the Giant Moose: Natural History in Early America*. Chicago: University of Chicago Press, 2009.

Dutra, Eliana R. de Freitas. "A nação nos livros: A biblioteca ideal na coleção Brasiliana." In *Política, Nação e Edição*, ed. Eliana R. de Freitas Dutra and Jean Yves Mollier, 299–314. São Paulo: AnnaBlume, 2006.

———. *Rebeldes literários da República: História e identidade nacional no Almanaque Brasileiro Garnier, 1903–1914*. Belo Horizonte: Editora UFMG, 2005.

Eakin, Marshall, and Paulo Almeida. *Envisioning Brazil: A guide to Brazilian Studies in the United States*. Madison: University of Wisconsin Press, 2005.

Faria, Luís de Castro. *Alberto Childe: O homem, a obra e a sua época*. Publicações Avulsas do Museu Nacional 60. Rio de Janeiro: Museu Nacional, 1970.

———. "Um sábio maranhense no Museu Nacional." In *Raimundo Lopes, dois estudos resgatados*, ed. Heloisa Maria Bertol Domingues and Alfredo Almeida, 7–13. Rio de Janeiro: Ouro sobre Azul, 2010.

Fausto, Bóris. *A Revolução de 30*. São Paulo: Brasiliense, 1970.

———. *História do Brasil*. 12th ed. São Paulo: Edusp, 2004.

Favret, Colin. "Jean-Henri Fabre: His Life, Experiences and Predisposition against Darwinism." *American Entomologist* 45:1 (1999): 38–48.

Ferreira, Luiz Fernando. "Lauro Travassos." *Cadernos de Saúde Pública* 5:4 (1989): 461–69.

Ferreira, Marieta de Moraes. "A reação republicana e a crise política nos anos 1920." *Estudos Históricos* 6:11 (1993): 9–23.

Ferri, Mário. "A botânica no Brasil." In *As ciências no Brasil*, ed. Fernando Azevedo, 2:149–200. São Paulo: Melhoramentos, 1956.

Figueiroa, Silvia. "Mundialização da ciência e respostas locais." *Asclepio* 50:2 (1998): 107–23.

Flechtmann, Carlos A. H., and Angelo L. T. Ottati. "*Tetanorhynchus leonardosi* (Mello-Leitão) (Orthoptera: Proscopiidae), nova praga em eucaliptos." *Anais da Sociedade Entomológica do Brasil* 26:3 (1997): 583–87.

Formigni, Mileide, and Hilton Silva, "Conservação ambiental e populações tradicionais: Uma contribuição bibliográfica da obra de A. Ruschi." *Boletim do Museu de Biologia Mello Leitão* 31 (2013): 50–75.

Foucault, Michel. *The Archaeology of Knowledge*. New York: Pantheon, 1982.

———. "Cuvier's Position in the History of Biology." *Critique of Anthropology* 4:13–14 (1979): 125–30.

———. *Discipline and Punish: The Birth of the Prison*. New York: Vintage, 1995.

———. *The Order of Things: An Archaeology of the Human Sciences*. New York: Vintage, 1994.

———. *Security, Territory, Population: Lectures at the Collège de France, 1977–1978*. New York: Picador, 2009.

———. *Society Must Be Defended: Lectures at the Collège de France, 1975–1976*. New York: Picador, 2003.

———. "Truth and Power." In *Power*, vol. 3 of *Essential Works of Foucault, 1954–1984*, ed. James D. Faubion, 111–33. New York: New Press, 2001.

Franco, José Luiz de Andrade, and José Augusto Drummond. "Alberto José Sampaio: Um botânico brasileiro e o seu programa de proteção à natureza." *Varia Historia* 21:33 (2005): 129–59.

———. "Cândido de Mello Leitão: As ciências biológicas e a valorização da natureza e da diversidade da vida." *História, Ciências, Saúde—Manguinhos* 14:4 (2006): 1265–90.

———. "Frederico Carlos Hoehne: A atualidade de um pioneiro no campo da proteção à natureza no Brasil." *Ambiente e sociedade* 8:1 (2006): 141–66.

———. *Proteção à natureza e identidade nacional no Brasil, anos 1920–1940*. Rio de Janeiro: Editora Fiocruz, 2009.

———. "Wilderness and the Brazilian Mind I: Nation and Nature in Brazil from the 1920s to the 1940s." *Environmental History* 13 (2008): 724–50.

———. "Wilderness and the Brazilian Mind II: The First Brazilian Conference on Natural Protection." *Environmental History* 14 (2009): 82–102.

Freitas, Marcus Vinicius de. *Charles Frederick Hartt, um naturalista no império de Pedro II*. Belo Horizonte: Editora UFMG, 2002.

Galton, Francis. *Hereditary Genius: An Inquiry into Its Laws and Consequences*. London: Macmillan, 1869.

———. *Natural Inheritance*. London: Macmillan, 1889.

Galvão, Elisandra. "A ciência vai ao cinema: Uma análise de filmes educativos e de divulgação científica do Instituto Nacional do Cinema Educativo (INCE)." MA thesis, Universidade Federal do Rio de Janeiro, 2004.

Gerbi, Antonello. *The Dispute of the New World: Tthe History of a Polemic*. Pittsburgh: University of Pittsburgh Press, 2010.

Gilioli, Renato de Souza. "Educação e cultura no rádio brasileiro: Concepções de radioescola em Roquette-Pinto." PhD diss., Universidade de São Paulo, 2008.

Glick, Thomas. "A Fundação Rockefeller e a emergência da genética no Brasil (1943–1960)." In *A recepção do darwinismo no Brasil*, ed. Heloisa Maria Bertol Domingues, Magali Romero Sá, and Thomas Glick, 145–74. Rio de Janeiro: Editora Fiocruz, 2003.

———. "Science and Independence in Latin America." *Hispanic American Historical Review* 71:2 (1991): 306–34.

Godoy, Solange. *O avô do tempo: Diário de um meterologista*. Rio de Janeiro: EMC, 2009.

Gomes, Angela de Castro. "Confronto e compromisso no processo de constitucionalização (1930–1935)." In *O Brasil Republicano*, ed. Bóris Fausto, 8–75. 2nd ed. Coleção História Geral da Civilização Brasileira 10. São Paulo: Difel, 1983.

———. *A República, a história e o IHGB*. Belo Horizonte: Argvmentvm, 2009.

Gomes, Angela de Castro, and Martha Abreu. "A nova 'Velha' República: Um pouco de história e historiografia." *Tempo* 13:26 (2009): 1–14.

Gomes, Maria do Carmo Andrade. *A pequena pátria: Políticas cartográficas em Minas Gerais*. São Paulo: AnnaBlume, 2015.

Gould, Stephen J. *Ever Since Darwin: Reflections on Natural History*. New York: W.W. Norton, 1977.

———. *The Structure of Evolutionary Theory*. 15th ed. Cambridge, MA: Harvard University Press, 2002.

Gouvea Filho, Pedro. "E. Roquette-Pinto, antropólogo e educador." *Revista Brasileira de Estudos Pedagógicos* 24:59 (1955): 31–57.

Grove, Richard. *Green Imperialism: Colonial Expansion, Tropical Island Edens and the Origins of Environmentalism, 1600–1860*. New York: Cambridge University Press, 1997.

Gualtieri, Regina. "O evolucionismo na produção científica do Museu Nacional do Rio de Janeiro (1876–1915)." In *A recepção do darwinismo no Brasil*, ed. Heloisa Maria Bertol Domingues, Magali Romero Sá, and Thomas Glick, 45–96. Rio de Janeiro: Editora Fiocruz, 2003.

Health, Christian, and D. Vorn Lehn. "Configuring Interactivity: Enhancing Engagement in Science Centres and Museums." *Social Studies of Science* 38:1 (2008): 63–91.

Heizer, Alda. "Considerações sobre a participação da América Latina nas grandes Exposições." *Revista da ANPLAC* 9:1 (2010): 15–28.

Hilton, Stanley. *Oswaldo Aranha: Uma biografia*. Rio de Janeiro: Objetiva, 1994.

Hobsbawm, Eric. *The Age of Extremes: A History of the World, 1914–1991*. New York: Vintage, 1996.

Hoffenberg, Peter. *An Empire on Display*. Berkeley: University of California Press, 2001.

Holloway, Thomas. *Immigrants on the Land: Coffee and Society in São Paulo, 1886–1934*. Chapel Hill: University of North Carolina Press, 1980.

Hurley, Patrick. "The Confirmation of Continental Drift." *Scientific American* 218 (1968): 52–64.

Iglésias, Francisco. *Trajetória política do Brasil*. São Paulo: Companhia das Letras, 1995.

Jacoby, Wolfgang. "Modern Concepts of Earth Dynamics Anticipated by Alfred Wegener in 1912." *Geology: Journal of the Geological Society of America* 9:1 (1981): 25–27.

Junghans, Miriam. "Emília Snethlage (1686–1929): Uma naturalista alemã na Amazônia." *História, Ciências, Saúde—Manguinhos* 15, suppl. (2008): 243–55.

Kohlstedt, Sally G. "A Better Crop of Boys and Girls: The School Gardening Movement, 1890–1920." *History of Education Quarterly* 48:1 (2008): 58–93.

———. "Nature, Not Books: Scientists and the Origins of the Nature-Study Movement in the 1890s." *Isis* 96:3 (2005): 324–52.

———. "Thoughts in Things: Modernity, History, and North American Museums." *Isis* 96:4 (2005): 586–601.

Koifman, Fábio, ed. *Presidentes do Brasil*. Rio de Janeiro: Cultura, Editora Rio, 2002.

Kropf, Simone. *Doença de Chagas, doença do Brasil: Ciência, saúde e nação 1909–1962*. Rio de Janeiro: Editora Fiocruz, 2009.

Kury, Adriano, and Renner Baptista. "Arachnological Papers published by Cândido Firmino de Mello-Leitão (Arachnida)." *Publicações Avulsas do Museu Nacional* 105. Rio de Janeiro: Museu Nacional, 2004.

Kury, Lorelai. *Comissão Científica do Império*. Rio de Janeiro: Andrea J. Editorial, 2009.

Laver, James. *Costume and Fashion: A Concise History*. Oxford: Oxford University Press, 1983.

Lemos, Renato. "Política externa." In *Getúlio Vargas e seu tempo*, ed. Raul Silva Mendes, Paulo Capachuz, and Sergio Lamarão, 75–82. Rio de Janeiro: BNDES, 2004.

Lenharo, Alcir. *A sacralização da política*. São Paulo: Papirus, 1986.

Lima, Antônio. *Um grande cerco de paz*. Petropolis: Vozes, 1995.

Lima, Nísia Trindade. *Um sertão chamado Brasil*. Rio de Janeiro: Revan, 1999.

Lima, Nísia Trindade, and Dominichi Miranda de Sá, eds. *Catálogo da exposição Roquette-Pinto um brasiliano*. Rio de Janeiro: Fiocruz, Museu da Vida, Casa da Ciência da UFRJ, Radio MEC, 2004.

———. "Roquette-Pinto e sua geração na república das letras e da ciência." In *Antropologia brasiliana: Ciência e educação na obra de Edgard Roquette-Pinto*, ed. Lima and Sá, 57–63. Rio de Janeiro: Editora Fiocruz, Belo Horizonte: Editora UFMG, 2008.

Lisboa, Karen. *A nova Atlântida de Spix e Martius: Natureza e civilização na Viagem pelo Brasil*. São Paulo: Hucitec, 1997.

Lopes, Margareth. "A formação dos museus nacionais e a América Latina independente." *Anais do Museu Histórico Nacional* 30:1 (1998): 121–33.

———. *"Cooperacão científica na América latina." Interciencias* 25:5 (2000): 228–33.

———. *O Brasil descobre a pesquisa científica*. São Paulo: Hucitec, 1997.

Loureiro, José Mauro. "Museu de ciência, divulgação científica e hegemonia." *Ciência da Informação* 32:1 (2003): 88–95.

Luca, Tania de. *A revista do Brasil: Um diagnóstico para a (N)ação*. São Paulo: Editora Unesp, 1999.

Lussana, Fiamma. "Cinema 'Educatore': Luciano de Feo direttore dell'Istituto Luce." *Studi Storici* 56:4 (2015): 935–62.

Lyra, Maria de Lourdes Viana. *A utopia do poderoso império*. Rio de Janeiro: Sette Letras, 1994.

Malerba, Jurandir. *A Corte no exílio*. São Paulo: Companhia das Letras, 2001.

Marandino, Martha. "Museus de ciências como espaços de educação." In *Museus: Dos gabinetes de curiosidades à museologia moderna*, ed. Betânia Gonçalves Figueiredo and Diana Gonçalves Vidal, 165–76. Belo Horizonte: Argvmentvm, 2005.

Marcovitch, Jacques. *Pioneiros e empreendedores: A saga do desenvolvimento no Brasil*. São Paulo: Edusp, 2005.

Marson, Adalberto. *A ideologia nacionalista em Alberto Torres*. São Paulo: Duas Cidades, 1979.

Martini, Augusto Jeronimo. "O plantador de eucaliptos: A questão da preservação florestal no Brasil e o resgate documental do legado de Edmundo Navarro de Andrade." MA thesis, Universidade de São Paulo, 2004.

Massarani, Luisa, and Nísia Trindade Lima. "Projeto Memória da Rádio Sociedade." Rio de Janeiro: Fiocruz. www.fiocruz.br/radiosociedade/cgi/cgilua.exe/sys/start.htm?sid=35, accessed July 27, 2015.

Matos, Everton Coelho. "Brasil e Uruguai: Uma dívida que virou ponte." Coursework, Pontifícia Universidade Católica do Rio Grande do Sul, Uruguaiana, 2008.

Mattos, Hebe. "Vida política." In *História do Brasil Nação: A abertura para o mundo*, ed. Lilia Schwarcz, 85–132. São Paulo: Objetiva, 2013.

Mattos, Hebe, and Ana Lugão Rios. *Memórias do cativeiro: Família, trabalho e cidadania no pós-abolição.* São Paulo: Civilização Brasileira, 2005.

Mayr, Ernst. *The Growth of Biological Thought: Diversity, Evolution and Inheritance.* 11th ed. Cambridge, MA: Harvard University Press, 2000.

———. *What Makes Biology Unique? Considerations on the Autonomy of a Scientific Discipline.* New York: Cambridge University Press, 2004.

McCann, Frank, Jr. *The Brazilian-American Alliance, 1937–1945.* Princeton: Princeton University Press, 1973.

McCook, Stuart. "Crônica de uma praga anunciada: Epidemias agrícolas e história ambiental do café nas Américas." *Varia Historia* 24:39 (2008): 87–112.

———. *States of Nature: Science, Agriculture and Environment in the Spanish Caribbean, 1760–1940.* Austin: University of Texas Press, 2002.

Meneses, Ulpiano. "A exposição museológica e o conhecimento histórico." In *Museus: Dos gabinetes de curiosidades à museologia moderna,* ed. Betânia Gonçalves Figueiredo and Diana Gonçalves Vidal, 15–84. Belo Horizonte: Argvmentvm, 2005.

Mitman, Gregg. *The State of Nature: Ecology, Community and American Social Thought, 1900–1950.* Chicago: University of Chicago Press, 1992.

Monarcha, Carlos. *Anísio Teixeira: Obra de uma vida.* Rio de Janeiro: DP&A, 2001.

———. *Brasil arcaico, Escola Nova: Ciência, técnica e utopia nos anos 1920–1930.* São Paulo: Ed. Unesp, 2009.

Moraes, José Damiro de. "Signatárias do Manifesto de 1932: Trajetórias e dilemas." PhD diss., Universidade Estadual de Campinas, 2007.

Moraes, Maria Célia. "Educação e política nos anos 30: A presença de Francisco Campos." *Revista Brasileira de Estudos Pedagógicos* 73:174 (1992): 201–321.

Morais, Rita de Cássia de Jesus. "Nos verdes campos da ciência: A trajetória acadêmica do médico e botanista brasileiro Francisco de Freire Alemão." MA thesis, Fundação Oswaldo Cruz, Rio de Janeiro, 2005.

Moreira, Ildeu de Castro, and Luisa Massarani. "A divulgação científica no Rio de Janeiro: Algumas reflexões sobre a década de 20." *História, Ciências, Saúde—Manguinhos* 7:3 (2000/2001): 327–51.

Moreira, Ildeu de Castro, Luisa Massarari, and Jayme Aranha. "Roquette-Pinto e a divulgação científica." In *Antropologia brasiliana: Ciência e educação na obra de Edgard Roquette-Pinto,* ed. Nísia Trindade Lima and Dominichi Miranda de Sá, 247–70. Rio de Janeiro: Editora Fiocruz; Belo Horizonte: Editora UFMG, 2008.

Motta, Marly Silva da. *A nação faz cem anos.* Rio de Janeiro: CPDOC/FGV, 1992.

Neves, Livia Lopes. "Pensamento da América: Intelectualidade e Estado Novo em um projeto comungado (1941–1945)." MA thesis, Universidade Federal de Santa Catarina, Florianópolis, 2013.

Nyhart, Lynn K. "Natural History and the 'New' Biology." In *Cultures of Natural History*, ed. Nicholas Jardine, James A. Secord, and Emma C. Spary, 426–43. Cambridge: Cambridge University Press, 1996.

O Museu Nacional. São Paulo: Banco Safra, 2007.

Otovo, Okezi T. "From Mãe Preta to Mãe Desamparada: Maternity and Public Health in Post-Abolition Bahia." *Luso-Brazilian Review* 48:2 (2011): 164–91.

Pádua, José Augusto. *Um sopro de destruição: Pensamento político e crítica ambiental no Brasil escravista, 1786–1888*. Rio de Janeiro: Jorge Zahar, 2004.

Pallares-Burke, Maria Lúcia. *Gilberto Freyre: Um vitoriano dos trópicos*. São Paulo: Editora Unesp, 2005.

Pandolfi, Dulce. "Os anos 1930: As incertezas do regime." In *O Brasil Republicano: O tempo do nacional-estatismo*, ed. Jorge Ferreira and Lucilia Delgado, 13–37. Rio de Janeiro: Civilização Brasileira, 2003.

Pauly, Philip. *Biologists and the Promise of American Life*. Princeton: Princeton University Press, 2002.

Peixoto, Ana Maria Cassasanta. "A escola no projeto de construção do Brasil moderno: A reforma Francisco Campos em Minas Gerais." *Educação em Revista* 16 (1992): 12–17.

Pereira, Osny Duarte. *Direito florestal brasileiro*. Rio de Janeiro: Borsoi, 1950.

Pesavento, Sandra. "Imagens da nação, do progresso e da tecnologia: A Exposição Universal de 1876." *Anais do Museu Paulista* 2:1 (1994): 151–67.

Peters, James. *Check List of Birds of the World*. Cambridge, MA: Harvard University Press, 1931.

Pinheiro, Letícia. *Política externa brasileira*. Rio de Janeiro: Zahar, 2004.

Pinheiro Filho, Fernando A. "A invenção da ordem: Intelectuais católicos no Brasil." *Tempo Social* 19:1 (2007): 33–49.

Pinto, Olivério M. de Oliveira. "A zoologia no Brasil." In *As ciências no Brasil*, ed. Fernando Azevedo, 2:93–148. São Paulo: Melhoramentos, 1956.

Podgorny, Irina. "El ordenamiento de la naturaleza a través de los museos y de la ciencia en la Confederacíon Argentina." *Quipu* 12:2 (1999): 167–86.

Pombal Junior, José. "Nota biográfica sobre Alípio de Miranda Ribeiro." *Revista Brasileira de Zoologia* 19:3 (2002): 935–39.

Pontes, Heloisa. "Retratos do Brasil: Editores, editoras e 'Coleções Brasiliana' nas décadas de 30, 40 e 50." In *História das Ciências Sociais no Brasil*, ed. Sérgio Miceli, 359–409. São Paulo: Vértice, 1989.

Prestes, Maria Elice. *A investigação da natureza no Brasil colônia*. São Paulo: Annablume, 2000.

Pruna, Pedro. "National Science in a Colonial Context." *Isis* 85:3 (1994): 412–26.

Rangel, Marcio Ferreira. "Um entomólogo chamado Costa Lima: A construção de um saber e a consolidação de um patrimônio científico." PhD diss., Fundação Oswaldo Cruz, Rio de Janeiro, 2006.

Rebouças, Márcia. "Pelo resgate da memória documental das ciências e da agricultura: O acervo do Instituto Biológico de São Paulo." *História, Ciências, Saúde—Manguinhos* 13:4 (2006): 995–1005.

Resende, Maria Efigênia Lage de. "O processo politico no Primeira República e o liberalismo oligárquico." In *O Brasil Republicano: O tempo do liberalismo excludente*, ed. Jorge Ferreira and Lucília Delgado, 89–120. Rio de Janeiro: Civilização Brasileira, 2003.

Rezende, Livia. "Crafting the Nation: Brazilian 'Civilised Exoticism' at World Exhibitions." In *Design and Craft: A History of Convergences and Divergences*, ed. Javier Gimeno and Fredie Floré, 136–39. Brussels: Universa, 2010.

———. "Designing the State at Brazil's Independence Centennial International Exhibition." In *Design Frontiers*, ed. Priscila Farias and Paul Atkinson, 79–89. Mexico City: Editorial Designeo, 2014.

Rocha, Heloisa. "Prescrevendo regras de bem viver: Cultura escolar e racionalidade científica." *Cadernos Cedes* 20:52 (2000): 55–73.

Rodrigues, Cândido Moreira. *A Ordem: Uma revista de intelectuais católicos*. São Paulo: Fapesp, 2005.

Rodrigues, José Honório, and Ricardo Seitenfus. *Uma história diplomática do Brasil, 1531–1945*. Rio de Janeiro: Civilização Brasileira, 1995.

Roquette-Pinto, Vera. "Roquette-Pinto, o rádio e o cinema educativos." *Revista USP* 56 (2002–3): 10–15.

Ruschi, André. "Augusto Ruschi: Feitos e legados." www.augustoruschi.com.br, accessed May 5, 2015.

Sá, Dominichi Miranda de. *A ciência como profissão: Médicos, bacharéis e cientistas no Brasil*. Rio de Janeiro: Editora Fiocruz, 2006.

Sá, Dominichi Miranda de, Magali Romero Sá, and Nísia Trindade Lima. "Telégrafos e inventário do território no Brasil: As atividades científicas da Comissão Rondon (1907–1915)." *História, Ciências, Saúde—Manguinhos* 15:3 (2008): 779–810.

Saldaña, Juan José, and Consuelo Cuevas. "La invención en Mexico de la investigación científica professional." *Quipu* 12:3 (1999): 309–22.

Sanjad, Nelson. "Da 'abominável profissão de vampiros': Emílio Goeldi e os mosquitos no Pará (1905)." *História, Ciências, Saúde—Manguinhos* 10:1 (2003): 85–111.

———. *Emílio Goeldi (1859–1917): A ventura de um naturalista entre a Europa e o Brasil.* Rio de Janeiro: EMC Edições, 2009.

Santos, Luís Claudio Vilafañe. *O evangelho do Barão.* São Paulo: Editora Unesp, 2012.

Santos, Maria Cristina Ferreira dos. "A biologia de Cândido de Mello Leitão e a História Natural de Waldemiro Potsch: Professores, autores e livros didáticos, conhecimento e poder em disputa na constituição da biologia escolar." PhD diss., Universidade Federal Fluminense, Niterói, 2013.

Schvarzman, Sheila. *Humberto Mauro e as imagens do Brasil.* São Paulo: Editora Unesp, 2004.

Schwarcz, Lilia. *O espetáculo das raças: Cientistas, instituições e questão racial no Brasil, 1870–1930.* São Paulo: Companhia das Letras, 1993.

———. "Os trópicos como espetáculo." In *Galerias del Progreso*, ed. Beatriz González and Jens Andermann, 195–220. Buenos Aires: Beatriz Viterbo Editora, 2006.

Schwartzman, Simon. *Um espaço para a ciência.* Brasília: MCT, 2001.

Schwartzman, Simon, Helena Bomeny, and Vanda Costa. *Tempos de Capanema.* Rio de Janeiro: Paz e Terra, 2000.

Scott, James C. *Seeing Like a State: How Certain Schemes to Improve the Human Condition Have Failed.* New Haven: Yale University Press, 1998.

Sevcenko, Nicolau. "A capital irradiante: Técnica, ritmos e ritos do Rio." In *História da vida privada no Brasil: República, da Belle Époque à era do rádio*, ed. Nicolau Sevcenko, 513–619. São Paulo: Companhia das Letras, 1998.

Seyferth, Giralda. "Roquette-Pinto e o debate sobre raça e imigração no Brasil." In *Antropologia brasiliana: Ciência e educação na obra de Edgard Roquette-Pinto*, ed. Nísia Trindade Lima and Dominique Miranda de Sá, 147–77. Rio de Janeiro: Editora Fiocruz; Belo Horizonte: Editora UFMG, 2008.

Sheets-Pyenson, Susan. *Cathedrals of Science: The Development of Colonial Natural History Museums during the Late Nineteenth Century.* Kingston: McGill-Queen's University Press, 1988.

Silva, André Felipe Cândido da. "A campanha contra a broca-do-café em São Paulo (1924–1927)." *História, Ciências, Saúde—Manguinhos* 13:4 (2006): 957–93.

Silva, João Ítalo de Oliveira e. "Por uma eugenia latino-americana: Victor Delfino e Renato Kehl." MA thesis, Universidade Federal de Minas Gerais, Brazil, 2008.

Soluri, John. *Banana Cultures: Agriculture, Consumption, and Environmental Change in Honduras and the United States.* Austin: University of Texas Press, 2005.

Sousa, Lia Gomes Pinto de, Mariana Moraes de Oliveira Sombrio, and Margareth Lopes. "Para ler Bertha Lutz." *Cadernos Pagu* 24 (2005): 315–25.

Sousa, Vanderlei Sebastião de. *Edgard Roquette-Pinto e o retrato antropológico brasileiro, 1905–1935.* Natal: Editora UFRN; Rio de Janeiro: Editora FGV, 2015.

Souza, Gilda de Mello e. *O espírito das roupas.* São Paulo: Companhia das Letras, 1987.

Starling, Heloisa, Ligia Germano, and Rita de Cássia Marques, eds. *Medicina, história em exame.* Belo Horizonte: Editora UFMG, 2011.

Stepan, Nancy. *Beginnings of Brazilian Science: Oswaldo Cruz, Medical Research and Policy, 1890–1920.* New York: Science History Publications, 1976.

Teive, Hélio A. G., Sérgio M. De Almeida, and Lineu César Werneck. "The Brazilian Contribution to the Study of Neurocysticercosis: Moses and Lange's Role in Cerebrospinal Fluid Diagnosis." *Arquivos de Neuro-Psiquiatria* 64:2B (2006): 534–37.

Thielen, Eduardo Vilela, and Ricardo Augusto dos Santos. "Belisário Penna: Notas fotobiográficas." *História, Ciências, Saúde—Manguinhos* 9:2 (2002): 387–404.

Toledo, Maria Rita de Almeida. "Coleção 'Atualidades Pedagógicas': Do projeto político ao projeto editorial (1931–1981)." PhD diss., Universidade de São Paulo, 2001.

———. "O projeto político-cultural da Coleção Atualidades Pedagógicas." In *Política, nação e edição: O lugar dos impressos na construção da vida política*, ed. Eliana R. de Freitas Dutra and Jean Yves Mollier, 335–50. São Paulo: AnnaBlume, 2006.

Tort, Patrick. *Fabre, le miroir aux insectes.* Paris: Éditions Vuibert, 2002.

Urban, Teresa. *Saudade do matão: Relembrando a história da conservação da natureza no Brasil.* Curitiba: Editora UFPR, Fundação O Boticário, Fundação MacArthur, 1998.

Valente, Maria Esther, Sibele Cazelli, and Fátima Alves. "Museus, ciência e educação: Novos desafios." *História, Ciências, Saúde—Manguinhos* 12, suppl. (2005): 183–203.

Valero, Mercedez, and Armando García. "Ciencia y colecionismo en Cuba en el siglo XIX." *Asclepio* 51:1 (1999): 205–26.

Veiga, Cynthia Greive. "Educação estética para o povo." In *500 anos de educação no Brasil*, ed. Eliana Marta et al., 399–422. Belo Horizonte: Autêntica, 2000.

Vergara, Moema. "Ciência e história no Relatório da Comissão Exploradora do Planalto Central na Primeira República." *História, Ciências, Saúde—Manguinhos* 13:4 (2006): 909–25.

Vidal, Diana Gonçalves. "Manifiesto de los pioneros de la educación nueva." *Transatlántica de Educación* 5 (2008): 50–63.

———. *O exercício disciplinado do olhar: Livros, leituras e práticas de formação docente no Instituto de Educação do Distrito Federal (1932–1937)*. Bragança Paulista: Editora USF, 2001.

Viviani, Luciana. *A biologia necessária, formação de professoras e escola normal*. Belo Horizonte: Argvmentvm, 2007.

Wegener, Alfred. "The Origins of the Continents." 1915. Translated by Wolfgang Jacoby. *Journal of Geodynamics* 32:1–2 (2001): 20–63.

West, David A. *Fritz Müller: A Naturalist in Brazil*. Blacksburg: Pocahontas Press, 2003.

Williams, Daryle. *Culture Wars in Brazil: The First Vargas Regime, 1930–1945*. Durham: Duke University Press, 2001.

Zavaro, Thereza. *Os diretores do Museu Nacional/UFRJ*. Rio de Janeiro: Museu Nacional, 2008.

INDEX

Note: Page numbers in *italics* indicate illustrations.

ABOUT THE AUTHOR

Regina Horta Duarte is Full Professor of History at the Universidade Federal de Minas Gerais, Brazil. She is the author of the books *A Imagem Rebelde: A trajetória libertária de Avelino Fóscolo* (1991), *Noites Circenses: Espetáculos de circo e teatro em Minas Gerais, Brasil* (1995), *História e Natureza* (2005), and *A Biologia Militante* (2010). She also has published articles in several Brazilian journals and in international journals such as *Latin American Research Review*, *Environmental and History*, *Journal of Latin American Studies*, *Luso-Brazilian Review*, and *ISIS*.

She is a founding member of the Sociedad Latinoamericana y Caribeña de Historia Ambiental, and was the founding editor in chief of the journal *Historia Ambiental Latinoamericana y Caribeña*, from 2011 to 2014.

Her current research interests focus on the environmental history of zoos in Latin America in the first half of the twentieth century.

ABOUT THE TRANSLATOR

Diane Grosklaus Whitty is a full-time translator specializing in Brazilian studies.